"Carl Sagan is the most effective and popular advocate of the wonders of science in the United States."
—*The New York Times Books Review*

"He not only can make complex scientific matters understandable to the general reader, but does so in entrancing ways as well."
—Associated Press

"Sagan can write about anything...and seem as if he learned what he knows while playing in the sandbox."
—*The New York Times*

"Sagan has a love affair going with the universe....You cannot come away unmoved from an encounter with him. He is full of intensity, of fascination, of courtesy, and of a lust for accuracy and truth....Sagan blows through my head like a tornado, filling me with questions and with joy."
—*The Detroit News*

"Sagan overwhelms us, overpowering our senses....[He] sweeps us off our simple planet earth and into interplanetary space."
—*The San Diego Union*

Please turn the page
for more reviews.. .

"Devastating, balanced, unforgettable."
—*Science*

"He has the confidence and ability to range widely beyond his specialties, an infectious enthusiasm for ideas, delight in the 'romance of science' and optimism for the future....Sagan has probably done more than anyone else to make wonders of science readily available to the intelligent lay readers of this country."

—*The Pittsburgh Press*

"Throughout, Sagan's writing is eloquent and refreshing, his subjects wide-ranging and stimulating. He easily communicates the careful reason, the lively imagination, and the contagious enthusiasm that have made him the nation's best-known scientist."

—*The Charlotte Observer*

"The love and enthusiasm he exudes in his writings is infectious. He goads us to greater understanding by stimulating our intelligence....Sagan dazzles."

—*American Way*

"Intellectually omnivorous, utterly understandable...A brilliant book!"

—*Vogue*

BROCA'S BRAIN

REFLECTIONS ON THE ROMANCE OF SCIENCE

CARL SAGAN

BALLANTINE BOOKS • NEW YORK

 Published in the United States of America by Ballantine Books, a division of Random House, Inc., New York, and simultaneously in Canada by Random House of Canada Limited, Toronto.

Portions of this work have previously appeared in *American Scholar, Atlantic Monthly, Book Digest, Holiday, Mercury, Natural History, New Republic, New York Times Magazine, Physics Today, Playboy, Scientific American, Smithsonian Magazine, TV Guide,* and *Vogue* (British).
Grateful acknowledgment is made to the following for permission to reprint previously published material:
Cornell University Press: "An Analysis of 'Worlds in Collision' " by Carl Sagan, in *Scientists Confront Velikovsky,* edited by D. Goldsmith. Copyright © 1976 by Cornell University Press.
Encyclopedia Americana: "UFO's." Copyright © 1975 by Americana Corporation.
Field Enterprises: "The Climates of Planets" in *Science Year* 1975. Copyright © 1975 by Field Enterprises Educational Corporation.
William Morrow & Company, Inc.: Excerpts from *The Planets* by Diane Ackerman. Copyright © 1975, 1976 by Diane Ackerman. Reprinted by permission of William Morrow & Company, Inc.

Library of Congress Catalog Card Number: 78-21810

ISBN 0-345-33689-5

This edition published by arrangement with Random House, Inc.

Manufactured in the United States of America

First Ballantine Books Edition: October 1980
Twentieth Printing: October 1993

"The best nonfiction bet of the year."
—*The Denver Post*

"Every generation needs one scientist of impeccable credentials who is capable of explaining, in reasonably clear English and without sounding patronizing, what the hell is going on in our labs and think-tanks. At the moment, the chair is occupied (and very ably) by Carl Sagan, whose *Dragons of Eden* has already garnered a Pulitzer Prize."
—*The Houston Post*

"In the hands of an extraordinary scientist like Carl Sagan, the presentation of science as it really is becomes the exhilarating adventure it ought to be."
—*Worcester Telegram*

"Articulate, entertaining, and occasionally profound, he does what few other respected scientists can or have bothered to do—he gives science to the people. BROCA'S BRAIN is a marvelous gift."
—*The Providence Journal*

"Sagan is enjoying himself.... The puzzles and enigmas of the universe are unending. It is a scientist's delight in them that runs through all the parts of BROCA'S BRAIN."
—*Washington Star*

Also by Carl Sagan
Published by Ballantine Books:

THE DRAGONS OF EDEN

MURMURS OF EARTH

COSMOS

SHADOWS OF FORGOTTEN ANCESTORS
(with Ann Druyan)

To Rachel and Samuel Sagan, my parents,
who introduced me to the joys of understanding
the world, with gratitude and admiration and love

ACKNOWLEDGMENTS

FOR DISCUSSION on specific points I am grateful to a number of friends, correspondents and colleagues, including Diane Ackerman, D. W. G. Arthur, James Bakalar, Richard Berendzen, Norman Bloom, C. Chandrasekhar, Clark Chapman, Sidney Coleman, Yves Coppens, Judy-Lynn Del Rey, Frank Drake, Stuart Edelstein, Paul Fox, D. Carleton Gajdusek, Owen Gingerich, Thomas Gold, J. Richard Gott III, Steven J. Gould, Lester Grinspoon, Stanislav Grof, J. U. Gunter, Robert Horvitz, James W. Kalat, B. Gentry Lee, Jack Lewis, Marvin Minsky, David Morrison, Philip Morrison, Bruce Murray, Phileo Nash, Tobias Owen, James Pollack, James Randi, E. E. Salpeter, Stuart Shapiro, Gunther Stent, O. B. Toon, Joseph Veverka, E. A. Whitaker and A. Thomas Young.

This book owes much, in all stages of production, to the dedicated and competent efforts of Susan Lang, Carol Lane, and, particularly, my executive assistant, Shirley Arden.

I am especially grateful to Ann Druyan and Steven Soter for generous encouragement and stimulating commentary on a great many of the subjects of this book. Ann has made essential contributions to most chapters and to the title; my debt to her is very great.

PART I

SCIENCE AND HUMAN CONCERN

CHAPTER 1

BROCA'S BRAIN

◆◆◆

"They were apes only yesterday.
Give them time."
"Once an ape—always an ape." . . .
"No, it will be different. . . . Come back here in
an age or so and you shall see. . . ."

The gods, discussing the Earth, in the motion
picture version of H. G. Wells' *The Man Who
Could Work Miracles* (1936)

IT WAS A MUSEUM, in a way like any other, this Musée de l'Homme, Museum of Man, situated on a pleasant eminence with, from the restaurant plaza in back, a splendid view of the Eiffel Tower. We were there to talk with Yves Coppens, the able associate director of the museum and a distinguished paleoanthropologist. Coppens had studied the ancestors of mankind, their fossils being found in Olduvai Gorge and Lake Turkana, in Kenya and Tanzania and Ethiopia. Two million years ago there were four-foot-high creatures, whom we call *Homo habilis,* living in East Africa, shearing and chipping and flaking stone tools, perhaps building simple dwellings, their brains in the course of a spectacular enlargement that would lead one day—to us.

Institutions of this sort have a public and a private side. The public side includes the exhibits in ethnography, say, or cultural anthropology: the costumes of the Mongols, or bark cloths painted by Native Americans, some perhaps prepared especially for sale to *voyageurs* and enterprising French anthropologists. But in the innards of the place there are other things: people engaged in the construction of exhibits; vast storerooms of items inappropriate, because of subject matter or space, for general exhibition; and areas for research. We were led through a warren of dark, musty rooms, ranging from cubicles to rotundas. Research materials overflowed into the corridors: a reconstruction of a Paleolithic cave floor, showing where the antelope bones had been thrown after eating. Priapic wooden statuary from Melanesia. Delicately painted eating utensils. Grotesque ceremonial masks. Assagai-like throwing spears from Oceania. A tattered poster of a steatopygous woman from Africa. A dank and gloomy storeroom filled to the rafters with gourd woodwinds, skin drums, reed panpipes and innumerable other reminders of the indomitable human urge to make music.

Here and there could be found a few people actually engaged in research, their sallow and deferential demeanors contrasting starkly with the hearty bilingual competence of Coppens. Most of the rooms were evidently used for storage of anthropological items, collected from decades to more than a century ago. You had the sense of a museum of the second order, in which were stored not so much materials that might be of interest as materials that had once been of interest. You could feel the presence of nineteenth-century museum directors engaged, in their frock coats, in *goniométrie* and *craniologie*, busily collecting and measuring everything, in the pious hope that mere quantification would lead to understanding.

But there was another area of the museum still more remote, a strange mix of active research and virtually abandoned cabinets and shelves. A reconstructed and

CONTENTS

IV THE FUTURE

V ULTIMATE QUESTIONS

INTRODUCTION

WE LIVE in an extraordinary age. These are times of stunning changes in social organization, economic well-being, moral and ethical precepts, philosophical and religious perspectives, and human self-knowledge, as well as in our understanding of that vast universe in which we are imbedded like a grain of sand in a cosmic ocean. As long as there have been human beings, we have posed the deep and fundamental questions, which evoke wonder and stir us into at least a tentative and trembling awareness, questions on the origins of consciousness; life on our planet; the beginnings of the Earth; the formation of the Sun; the possibility of intelligent beings somewhere up there in the depths of the sky; as well as, the grandest inquiry of all—on the advent, nature and ultimate destiny of the universe. For all but the last instant of human history these issues have been the exclusive province of philosophers and poets, shamans and theologians. The diverse and mutually contradictory answers offered demonstrate that few of the proposed solutions have been correct. But today, as a result of knowledge painfully extracted from nature, through generations of careful thinking, observing and experimenting, we are on the verge of glimpsing at least preliminary answers to many of these questions.

There are a number of themes that weave through the structure of this book, appearing early, disappearing for a few chapters, and then resurfacing in a somewhat different context—including the joys and social consequences of the scientific endeavor; borderline or pop science; the not entirely different subject of religious doctrine; the exploration of the planets and the search for extraterrestrial life; and Albert Einstein, in the centenary of whose birth this book is published. Most of the chapters can be read independently, but the ideas have been presented in an order chosen with some care. As in some of my previous books, I have not hesitated to interject social, political or historical remarks where I thought they might be appropriate. The attention given to borderline science may seem curious to some readers. Practitioners of pop science were once called Paradoxers, a quaint nineteenth-century word used to describe those who invent elaborate and undemonstrated explanations for what science has understood rather well in simpler terms. We are today awash with Paradoxers. The usual practice of scientists is to ignore them, hoping they will go away. I thought it might be useful—or at least interesting—to examine the contentions and conceits of some Paradoxers a little more closely, and to connect and contrast their doctrines with other belief systems, both scientific and religious.

Both borderline science and many religions are motivated in part by a serious concern about the nature of the universe and our role in it, and for this reason merit our consideration and regard. In addition, I think it possible that many religions involve at their cores an attempt to come to grips with profound mysteries of our individual life histories, as described in the last chapter. But both in borderline science and in organized religion there is much that is specious or dangerous. While the practitioners of such doctrines often wish there were no criticisms to which they are expected to reply, skeptical scrutiny is the means, in both science and religion, by which deep insights can be winnowed from deep nonsense. I hope my critical remarks in these pages will be recognized as constructive in intent. The well-meaning contention that all ideas have equal merit seems to me little different from the disastrous contention that no ideas have any merit.

This book, then, is about the exploration of the universe and ourselves; that is, it is about science. The range of topics may seem very diverse—from a crystal articulating skeleton of an orangutan. A vast table covered with human skulls, each neatly indexed. A drawer full of femurs, piled in disarray, like the erasers in some school janitor's supply closet. A province dedicated to Neanderthal remains, including the first Neanderthal skull, reconstructed by Marcellin Boule, which I held cautiously in my hands. It felt lightweight and delicate, the sutures starkly visible, perhaps the first compelling piece of evidence that there once were creatures rather like us who became extinct, a disquieting hint that our species likewise might not survive forever. A tray filled with the teeth of many hominids, including the great nutcracker molars of *Australopithecus robustus*, a contemporary of *Homo habilis*. A collection of Cro-Magnon skull cases, stacked like cordwood, scrubbed white and in good order. These items were reasonable and in a way expected, the necessary shards of evidence for reconstructing something of the history of our ancestors and collateral relatives.

Deeper in the room were more macabre and more disturbing collections. Two shrunken heads reposing on a cabinet, sneering and grimacing, their leathery lips curled back to reveal rows of sharp, tiny teeth. Jar upon jar of human embryos and fetuses, pale white, bathed in a murky greenish fluid, each jar competently labeled. Most specimens were normal, but occasionally an anomaly could be glimpsed, a disconcerting teratology—Siamese twins joined at the sternum, say, or a fetus with two heads, the four eyes tightly shut.

There was more. An array of large cylindrical bottles containing, to my astonishment, perfectly preserved human heads. A red-mustachioed man, perhaps in his early twenties, originating, so the label said, from Nouvelle Calédonie. Perhaps he was a sailor who had jumped ship in the tropics only to be captured and executed, his head involuntarily drafted in the cause of science. Except he was not being studied; he was only being neglected, among the other severed heads. A sweet-faced and delicate little girl of perhaps four years, her pink coral earrings and necklace still perfectly preserved. Three infant heads, sharing the same bottle, perhaps as an economy measure. Men and women and children of both sexes and many races, decapitated, their heads shipped to France only to moulder—perhaps after some brief initial study—in the Musée de l'Homme. What, I wondered, must the loading of the crates of bottled heads have been like? Did the ship's officers speculate over coffee about what was down in the hold? Were the sailors heedless because the heads were, by and large, not those of white Europeans like themselves? Did they joke about their cargo to demonstrate some emotional distance from the little twinge of horror they privately permitted themselves to feel? When the collections arrived in Paris, were the scientists brisk and businesslike, giving orders to the draymen on the disposition of severed heads? Were they impatient to unseal the bottles and embrace the contents with calipers? Did the man responsible for this collection, whoever he might be, view it with unalloyed pride and zest?

And then in a still more remote corner of this wing of the museum was revealed a collection of gray, convoluted objects, stored in formalin to retard spoilage—shelf upon shelf of human brains. There must have been someone whose job it was to perform routine craniotomies on the cadavers of notables and extract their brains for the benefit of science. Here was the cerebrum of a European intellectual who had achieved momentary renown before fading into the obscurity of this dusty shelf. Here a brain of a convicted murderer. Doubtless the savants of earlier days had hoped there might be some anomaly, some telltale sign in the brain anatomy or cranial configuration of murderers. Perhaps they had hoped that murder was a matter of heredity and not society. Phrenology was a graceless nineteenth-century aberration. I could hear my friend Ann Druyan saying, "The people we starve and torture have an unsociable tendency to steal and murder. We think it's because their brows overhang." But the brains of murderers and savants—the remains of Albert Einstein's brain are floating wanly in a bottle in Wichita—are indistinguishable. It is, very probably, society and not heredity that makes criminals.

While scanning the collection amid such ruminations, my eye was caught by a label on one of the many low cylindrical bottles. I took the container from the shelf and examined it more closely. The label read *P. Broca.* In my hands was Broca's brain.

PAUL BROCA was a surgeon, a neurologist and an anthropologist, a major figure in the development of both medicine and anthropology in the mid-nineteenth century. He performed distinguished work on cancer pathology and the treatment of aneurisms, and made a landmark contribution to understanding the origins of aphasia—an impairment of the ability to articulate ideas. Broca was a brilliant and compassionate man. He was concerned with medical care for the poor. Under cover of darkness, at the risk of his own life, he successfully smuggled out of Paris in a horse-drawn cart 73 million francs, stuffed into carpetbags and hidden under potatoes, the treasury of the Assistance Publique which—he believed, at any rate—he was saving from pillage. He was the founder of modern brain surgery. He studied infant mortality. Toward the end of his career he was created a senator.

He loved, as one biographer said, mainly serenity and tolerance. In 1848 he founded a society of "freethinkers." Almost alone among French savants of the time, he was sympathetic to Charles Darwin's idea of evolution by natural selection. T. H. Huxley, "Darwin's Bulldog," remarked that the mere mention of Broca's name filled him with a sense of gratitude, and Broca was quoted as saying, "I would rather be a transformed ape than a degenerate son of Adam." For these and other views he was publicly denounced for "materialism" and, like Socrates, for corrupting the young. But he was made a senator nevertheless.

Earlier, Broca had encountered great difficulty in establishing a society of anthropology in France. The Minister of Public Instruction and the Prefect of Police believed that anthropology must, as the free pursuit of knowledge about human beings, be innately subversive to the state. When permission was at last and reluctantly granted for Broca to talk about science with eighteen colleagues, the Prefect of Police held Broca responsible personally for all that might be said in such meetings "against society, religion, or the government." Even so, the study of human beings was considered so dangerous that a police spy in plain clothes was assigned to attend all meetings, with the understanding that authorization to meet would be withdrawn immediately if the spy was offended by anything that was said. In these circumstances the Society of Anthropology of Paris gathered for the first time on May 19, 1859, the year of the publication of *The Origin of Species*. In subsequent meetings an enormous range of subjects was discussed —archaeology, mythology, physiology, anatomy, medicine, psychology, linguistics and history—and it is easy to imagine the police spy nodding off in the corner on many an occasion. Once, Broca related, the spy wished to take a small unauthorized walk and asked if he might leave without anything threatening to the state being said in his absence. "No, no, my friend," Broca responded. "You must not go for a walk: sit down and earn your pay." Not only the police but also the clergy opposed the development of anthropology in France, and in 1876 the Roman Catholic political party organized a major campaign against the teaching of the subject in the Anthropological Institute of Paris founded by Broca.

Paul Broca died in 1880, perhaps of the very sort of aneurism that he had studied so brilliantly. At the moment of his death he was working on a comprehensive study of brain anatomy. He had established the first professional societies, schools of research, and scientific journals of modern anthropology in France. His laboratory specimens became incorporated into what for many years was called the Musée Broca. Later it merged to become a part of the Musée de l'Homme.

It was Broca himself, whose brain I was cradling, who had established the macabre collection I had been contemplating. He had studied embryos and apes, and people of all races, measuring like mad in an effort to understand the nature of a human being. And despite the present appearance of the collection and my suspicions, he was not, at least by the standards of his time, more of a jingoist or a racist than most, and certainly not that standby of fiction and, more rarely, of fact: the cold, uncaring, dispassionate scientist, heedless of the human consequences of what he does. Broca very much cared.

In the *Revue d'Anthropologie* of 1880 there is a complete bibliography of Broca's writings. From the titles I could later glimpse something of the origins of the collection I had viewed: "On the Cranium and Brain of the Assassin Lemaire," "Presentation of the Brain of a Male Adult Gorilla," "On the Brain of the Assassin Prévost," "On the Supposed Heredity of Accidental Characteristics," "The Intelligence of Animals and the Rule of Humans," "The Order of the Primates: Anatomical Parallels between Men and Apes," "The Origin of the Art of Making Fire," "On Double Monsters," "Discussion on Microcephalics," "Prehistoric Trepanning," "On Two Cases of a Supernumerary Digit Developing at an Adult Age," "The Heads of Two New Caledonians" and "On the Skull of Dante Alighieri." I did not know the present resting place of the cranium of the author of *The Divine Comedy,* but the collection of brains and skulls and heads that surrounded me clearly began in the work of Paul Broca.

BROCA WAS a superb brain anatomist and made important investigations of the limbic region, earlier called the rhinencephalon (the "smell brain"), which we now know to be profoundly involved in human emotion. But Broca is today perhaps best known for his discovery of a small region in the third convolution of the left frontal lobe of the cerebral cortex, a region now known as Broca's area. Articulate speech, it turns out, as Broca inferred on only fragmentary evidence, is to an important extent localized in and controlled by Broca's area. It was one of the first discoveries of a separation of function between the left and right hemispheres of the brain. But most important, it was one of the first indications that specific brain functions exist in particular locales in the brain, that there is a connection between the anatomy of the brain and what the brain does, an activity sometimes described as "mind."

Ralph Holloway is a physical anthropologist at Columbia University whose laboratory I imagine must bear some resemblance to Broca's. Holloway makes rubber-latex casts of the insides of skulls of human and related beings, past and present, to attempt a reconstruction, from slight impressions on the interior of the cranium, of what the brain must have been like. Holloway believes that he can tell from a creature's cranium whether Broca's area is present, and he has found evidence of an emerging Broca's area in the brain of *Homo habilis* some two million years ago—just the time of the first constructions and the first tools. To this limited extent there is something to the phrenological vision. It is very plausible that human thought and industry went hand in hand with the development of articulate speech, and Broca's area may in a very real sense be one of the seats of our humanity, as well as a means for tracing our relationships with our ancestors on their way toward humanity.

And here was Broca's brain floating, in formalin and in fragments, before me. I could make out the limbic region which Broca had studied in others. I could see the convolutions on the neocortex. I could even make out the gray-white left frontal lobe in which Broca's own Broca's area resided, decaying and unnoticed, in a musty corner of a collection that Broca had himself begun.

It was difficult to hold Broca's brain without wondering whether in some sense Broca was still *in* there —his wit, his skeptical mien, his abrupt gesticulations when he talked, his quiet and sentimental moments. Might there be preserved in the configuration of neurons before me a recollection of the triumphant moment when he argued before the combined medical faculties (and his father, overflowing with pride) on the origins of aphasia? A dinner with his friend Victor Hugo? A stroll on a moonlit autumn evening, his wife holding a pretty parasol, along the Quai Voltaire and the Pont

Royal? Where do we go when we die? Is Paul Broca still there in his formalin-filled bottle? Perhaps the memory traces have decayed, although there is good evidence from modern brain investigations that a given memory is redundantly stored in many different places in the brain. Might it be possible at some future time, when neurophysiology has advanced substantially, to reconstruct the memories or insights of someone long dead? And would that be a good thing? It would be the ultimate breach of privacy. But it would also be a kind of practical immortality, because, especially for a man like Broca, our minds are clearly a major aspect of who we are.

From the character of this neglected storeroom in the Musée de l'Homme I had been ready to attribute to those who had assembled the collection—I had not known it was Broca at the time—a palpable sexism and racism and jingoism, a profound resistance to the idea of the relatedness of human beings and the other primates. And in part it was true. Broca was a humanist of the nineteenth century, but unable to shake the consuming prejudices, the human social diseases, of his time. He thought men superior to women, and whites superior to blacks. Even his conclusion that German brains were not significantly different from French ones was in rebuttal to a Teutonic claim of Gallic inferiority. But he concluded that there were deep connections in brain physiology between gorillas and men. Broca, the founder of a society of freethinkers in his youth, believed in the importance of untrammeled inquiry and had lived his life in pursuit of that aim. His falling short of these ideals shows that someone as unstinting in the free pursuit of knowledge as Broca could still be deflected by endemic and respectable bigotry. Society corrupts the best of us. It is a little unfair, I think, to criticize a person for not sharing the enlightenment of a later epoch, but it is also profoundly saddening that such prejudices were so extremely pervasive. The question raises nagging uncertainties about which of the conventional truths of our own age will be considered unforgivable bigotry by the next. One way to repay

Paul Broca for this lesson which he has inadvertently provided us is to challenge, deeply and seriously, our own most strongly held beliefs.

These forgotten jars and their grisly contents had been collected, at least partly, in a humanistic spirit; and perhaps, in some era of future advance in brain studies, they would prove useful once again. I would be interested in knowing a little more about the red-mustachioed man who had been, in part, returned to France from New Caledonia.

But the surroundings, the sense of a chamber of horrors, evoked unbidden other unsettling thoughts. At the very least, we feel in such a place a pang of sympathy for those—especially those who died young or in pain—who are in so unseemly a way thus memorialized. Cannibals in northwestern New Guinea employ stacked skulls for doorposts, and sometimes for lintels. Perhaps these are the most convenient building materials available, but the architects cannot be entirely unaware of the terror that their constructions evoke in unsuspecting passers-by. Skulls have been used by Hitler's SS, Hell's Angels, shamans, pirates, and even those who label bottles of iodine, in a conscious effort to elicit terror. And it makes perfectly good sense. If I find myself in a room filled with skulls, it is likely that there is someone nearby, perhaps a pack of hyenas, perhaps some gaunt and dedicated decapitator, whose occupation or hobby it is to collect skulls. Such fellows are almost certainly to be avoided, or, if possible, killed. The prickle of the hairs on the back of my neck, the increased heartbeat and pulse rate, that strange, clammy feeling are designed by evolution to make me fight or flee. Those who avoid decapitation leave more offspring. Experiencing such fears bestows an evolutionary advantage. Finding yourself in a room full of brains is still more horrifying, as if some unspeakable moral monster, armed with ghastly blades and scooping tools, were shuffling and drooling somewhere in the attics of the Musée de l'Homme.

But all depends, I think, on the purpose of the collection. If its objective is to find out, if it has acquired human parts *post mortem*—especially with the prior consent of those to whom the parts once belonged—then little harm has been done, and perhaps in the long run some significant human good. But I am not sure the scientists are entirely free of the motives of those New Guinea cannibals; are they not at least saying, "I live with these heads every day. They don't bother me. Why should *you* be so squeamish?"?

LEONARDO AND VESALIUS were reduced to bribery and stealth in order to perform the first systematic dissections of human beings in Europe, although there had been a flourishing and competent school of anatomy in ancient Greece. The first person to locate, on the basis of neuroanatomy, human intelligence in the head was Herophilus of Chalcedon, who flourished around 300 B.C. He was also the first to distinguish the motor from the sensory nerves, and performed the most thorough study of brain anatomy attempted until the Renaissance. Undoubtedly there were those who objected to his gruesome experimental predilections. There is a lurking fear, made explicit in the Faust legend, that some things are not "meant" to be known, that some inquiries are too dangerous for human beings to make. And in our own age, the development of nuclear weapons may, if we are unlucky or unwise, turn out to be a case of precisely this sort. But in the case of experiments on the brain, our fears are less intellectual. They run deeper into our evolutionary past. They call up images of the wild boars and highwaymen who would terrorize travelers and rural populations in ancient Greece, by Procrustean mutilation or other savagery, until some hero—Theseus or Hercules—would effortlessly dispatch them. These fears have served an adaptive and useful function in the past. But I believe they are mostly emotional baggage in the present. I was interested, as a scientist who has written about the brain, to find such revulsions hiding in me, to be revealed for my inspection in Broca's collection. These fears are worth fighting.

All inquiries carry with them some element of risk.

There is no guarantee that the universe will conform to our predispositions. But I do not see how we can deal with the universe—both the outside and the inside universe—without studying it. The best way to avoid abuses is for the populace in general to be scientifically literate, to understand the implications of such investigations. In exchange for freedom of inquiry, scientists are obliged to explain their work. If science is considered a closed priesthood, too difficult and arcane for the average person to understand, the dangers of abuse are greater. But if science is a topic of general interest and concern—if both its delights and its social consequences are discussed regularly and competently in the schools, the press, and at the dinner table—we have greatly improved our prospects for learning how the world really is and for improving both it and us. That is an idea, I sometimes fancy, that may be sitting there still, sluggish with formalin, in Broca's brain.

CHAPTER 2

CAN WE KNOW THE UNIVERSE? REFLECTIONS ON A GRAIN OF SALT

♦♦♦

Nothing is rich but the inexhaustible wealth
of nature. She shows us only surfaces,
but she is a million fathoms deep.

RALPH WALDO EMERSON

SCIENCE IS A WAY of thinking much more than it is a body of knowledge. Its goal is to find out how the world works, to seek what regularities there may be, to penetrate to the connections of things—from subnuclear particles, which may be the constituents of all matter, to living organisms, the human social community, and thence to the cosmos as a whole. Our intuition is by no means an infallible guide. Our perceptions may be distorted by training and prejudice or merely because of the limitations of our sense organs, which, of course, perceive directly but a small fraction of the phenomena of the world. Even so straightforward a question as whether in the absence of friction a pound of lead falls faster than a gram of fluff was answered incorrectly by

Aristotle and almost everyone else before the time of Galileo. Science is based on experiment, on a willingness to challenge old dogma, on an openness to see the universe as it really is. Accordingly, science sometimes requires courage—at the very least the courage to question the conventional wisdom.

Beyond this the main trick of science is to *really* think of something: the shape of clouds and their occasional sharp bottom edges at the same altitude everywhere in the sky; the formation of a dewdrop on a leaf; the origin of a name or a word—Shakespeare, say, or "philanthropic"; the reason for human social customs—the incest taboo, for example; how it is that a lens in sunlight can make paper burn; how a "walking stick" got to look so much like a twig; why the Moon seems to follow us as we walk; what prevents us from digging a hole down to the center of the Earth; what the definition is of "down" on a spherical Earth; how it is possible for the body to convert yesterday's lunch into today's muscle and sinew; or how far is up—does the universe go on forever, or if it does not, is there any meaning to the question of what lies on the other side? Some of these questions are pretty easy. Others, especially the last, are mysteries to which no one even today knows the answer. They are natural questions to ask. Every culture has posed such questions in one way or another. Almost always the proposed answers are in the nature of "Just So Stories," attempted explanations divorced from experiment, or even from careful comparative observations.

But the scientific cast of mind examines the world critically as if many alternative worlds might exist, as if other things might be here which are not. Then we are forced to ask why what we see is present and not something else. Why are the Sun and the Moon and the planets spheres? Why not pyramids, or cubes, or dodecahedra? Why not irregular, jumbly shapes? Why so symmetrical, worlds? If you spend any time spinning hypotheses, checking to see whether they make sense, whether they conform to what else we know, thinking of tests you can pose to substantiate or deflate your hypotheses, you will find yourself doing science. And as you come to practice this habit of thought more and more you will get better and better at it. To penetrate into the heart of the thing—even a little thing, a blade of grass, as Walt Whitman said—is to experience a kind of exhilaration that, it may be, only human beings of all the beings on this planet can feel. We are an intelligent species and the use of our intelligence quite properly gives us pleasure. In this respect the brain is like a muscle. When we think well, we feel good. Understanding is a kind of ecstasy.

But to what extent can we *really* know the universe around us? Sometimes this question is posed by people who hope the answer will be in the negative, who are fearful of a universe in which everything might one day be known. And sometimes we hear pronouncements from scientists who confidently state that everything worth knowing will soon be known—or even is already known—and who paint pictures of a Dionysian or Polynesian age in which the zest for intellectual discovery has withered, to be replaced by a kind of subdued languor, the lotus eaters drinking fermented coconut milk or some other mild hallucinogen. In addition to maligning both the Polynesians, who were intrepid explorers (and whose brief respite in paradise is now sadly ending), as well as the inducements to intellectual discovery provided by some hallucinogens, this contention turns out to be trivially mistaken.

Let us approach a much more modest question: not whether we can know the universe or the Milky Way Galaxy or a star or a world. Can we know, ultimately and in detail, a grain of salt? Consider one microgram of table salt, a speck just barely large enough for someone with keen eyesight to make out without a microscope. In that grain of salt there are about 10^{16} sodium and chlorine atoms. This is a 1 followed by 16 zeros, 10 million billion atoms. If we wish to know a grain of salt, we must know at least the three-dimensional positions of each of these atoms. (In fact, there is much more to be known—for example, the nature of the forces between the atoms—but we are making only a modest calculation.) Now, is this number more or less than the number of things which the brain can know?

How much *can* the brain know? There are perhaps 10^{11} neurons in the brain, the circuit elements and switches that are responsible in their electrical and chemical activity for the functioning of our minds. A typical brain neuron has perhaps a thousand little wires, called dendrites, which connect it with its fellows. If, as seems likely, every bit of information in the brain corresponds to one of these connections, the total number of things knowable by the brain is no more than 10^{14}, one hundred trillion. But this number is only one percent of the number of atoms in our speck of salt.

So in this sense the universe is intractable, astonishingly immune to any human attempt at full knowledge. We cannot on this level understand a grain of salt, much less the universe.

But let us look a little more deeply at our microgram of salt. Salt happens to be a crystal in which, except for defects in the structure of the crystal lattice, the position of every sodium and chlorine atom is predetermined. If we could shrink ourselves into this crystalline world, we would see rank upon rank of atoms in an ordered array, a regularly alternating structure—sodium, chlorine, sodium, chlorine, specifying the sheet of atoms we are standing on and all the sheets above us and below us. An absolutely pure crystal of salt could have the position of every atom specified by something like 10 bits of information.* This would not strain the information-carrying capacity of the brain.

If the universe had natural laws that governed its behavior to the same degree of regularity that determines a crystal of salt, then, of course, the universe would be knowable. Even if there were many such

* Chlorine is a deadly poison gas employed on European battlefields in World War I. Sodium is a corrosive metal which burns upon contact with water. Together they make a placid and unpoisonous material, table salt. Why each of these substances has the properties it does is a subject called chemistry, which requires more than 10 bits of information to understand.

laws, each of considerable complexity, human beings might have the capability to understand them all. Even if such knowledge exceeded the information-carrying capacity of the brain, we might store the additional information outside our bodies—in books, for example, or in computer memories—and still, in some sense, know the universe.

Human beings are, understandably, highly motivated to find regularities, natural laws. The search for rules, the only possible way to understand such a vast and complex universe, is called science. The universe forces those who live in it to understand it. Those creatures who find everyday experience a muddled jumble of events with no predictability, no regularity, are in grave peril. The universe belongs to those who, at least to some degree, have figured it out.

It is an astonishing fact that there *are* laws of nature, rules that summarize conveniently—not just qualitatively but quantitatively—how the world works. We might imagine a universe in which there are no such laws, in which the 10^{80} elementary particles that make up a universe like our own behave with utter and uncompromising abandon. To understand such a universe we would need a brain at least as massive as the universe. It seems unlikely that such a universe could have life and intelligence, because beings and brains require some degree of internal stability and order. But even if in a much more random universe there were such beings with an intelligence much greater than our own, there could not be much knowledge, passion or joy.

Fortunately for us, we live in a universe that has at least important parts that are knowable. Our common-sense experience and our evolutionary history have prepared us to understand something of the workaday world. When we go into other realms, however, common sense and ordinary intuition turn out to be highly unreliable guides. It is stunning that as we go close to the speed of light our mass increases indefinitely, we shrink toward zero thickness in the direction of motion, and time for us comes as near to stopping as we would like. Many people think that this is silly, and every week or two I get a letter from someone who complains to me about it. But it is a virtually certain consequence not just of experiment but also of Albert Einstein's brilliant analysis of space and time called the Special Theory of Relativity. It does not matter that these effects seem unreasonable to us. We are not in the habit of traveling close to the speed of light. The testimony of our common sense is suspect at high velocities.

Or consider an isolated molecule composed of two atoms shaped something like a dumbbell—a molecule of salt, it might be. Such a molecule rotates about an axis through the line connecting the two atoms. But in the world of quantum mechanics, the realm of the very small, not all orientations of our dumbbell molecule are possible. It might be that the molecule could be oriented in a horizontal position, say, or in a vertical position, but not at many angles in between. Some rotational positions are forbidden. Forbidden by what? By the laws of nature. The universe is built in such a way as to limit, or quantize, rotation. We do not experience this directly in everyday life; we would find it startling as well as awkward in sitting-up exercises, to find arms outstretched from the sides or pointed up to the skies permitted but many intermediate positions forbidden. We do not live in the world of the small, on the scale of 10^{-13} centimeters, in the realm where there are twelve zeros between the decimal place and the one. Our common-sense intuitions do not count. What does count is experiment—in this case observations from the far infrared spectra of molecules. They show molecular rotation to be quantized.

The idea that the world places restrictions on what humans might do is frustrating. Why *shouldn't* we be able to have intermediate rotational positions? Why *can't* we travel faster than the speed of light? But so far as we can tell, this is the way the universe is constructed. Such prohibitions not only press us toward a little humility; they also make the world more knowable. Every restriction corresponds to a law of nature, a regularization of the universe. The more restrictions there are on what matter and energy can do, the more knowledge human beings can attain. Whether in some sense the universe is ultimately knowable depends not only on how many natural laws there are that encompass widely divergent phenomena, but also on whether we have the openness and the intellectual capacity to understand such laws. Our formulations of the regularities of nature are surely dependent on how the brain is built, but also, and to a significant degree, on how the universe is built.

For myself, I like a universe that includes much that is unknown and, at the same time, much that is knowable. A universe in which everything is known would be static and dull, as boring as the heaven of some weak-minded theologians. A universe that is unknowable is no fit place for a thinking being. The ideal universe for us is one very much like the universe we inhabit. And I would guess that this is not really much of a coincidence.

CHAPTER 3

THAT WORLD WHICH BECKONS LIKE A LIBERATION

◆◆◆

To punish me for my contempt for authority,
Fate made me an authority myself.

EINSTEIN

ALBERT EINSTEIN was born in Ulm, Germany, in 1879, just a century ago. He is one of the small group of people in any epoch who remake the world through a special gift, a talent for perceiving old things in new ways, for posing deep challenges to conventional wisdom. For many decades he was a saintly and honored figure, the only scientist the average person could readily name. In part because of his scientific accomplishments, at least dimly grasped by the public; in part because of his courageous positions on social issues; and in part because of his benign personality, Einstein was admired and revered throughout the world. For scientifically inclined children of immigrant parents, or those growing up in the Depression, like me, the reverence accorded Einstein demonstrated that there were such people as scientists, that a scientific career might not be totally beyond hope. One major function he involuntarily served was as a scientific role model. Without Einstein, many of the young people who became scientists after 1920 might never have heard of the existence of the scientific enterprise. The logic behind Einstein's Special Theory of Relativity could have been developed a century earlier, but, although there were some premonitory insights by others, relativity had to wait for Einstein. Yet fundamentally the physics of special relativity is very simple, and many of the essential results can be derived from high school algebra and pondering a boat paddling upstream and downstream. Einstein's life was rich in genius and irony, a passion for the issues of his time, insights into education, the connection between science and politics, and was a demonstration that individuals can, after all, change the world.

As a child Einstein gave little indication of what was to come. "My parents," he recalled later, "were worried because I started to talk comparatively late, and they consulted the doctor because of it . . . I was at that time . . . certainly not younger than three." He was an indifferent student in elementary school, where he said the teachers reminded him of drill sergeants. In Einstein's youth, a bombastic nationalism and intellectual rigidity were the hallmarks of European education. He rebelled against the dull, mechanized methods of teaching. "I preferred to endure all sorts of punishment rather than learn to gabble by rote." Einstein was always to detest rigid disciplinarians, in education, in science and in politics.

At five he was stirred by the mystery of a compass. And, he later wrote, "at the age of 12 I experienced a second wonder of a totally different nature in a little book dealing with Euclidean plane geometry. . . . Here were assertions, as for example the intersection of the three altitudes of a triangle in one point, which—though by no means evident—could nevertheless be proved with such certainty that any doubt appeared to be out of the question. This lucidity and certainty made an indescribable impression upon me." Formal schooling provided only a tedious interruption to such contemplations. Einstein wrote of his self-education: "At the age of 12 to 16 I familiarized myself with the elements of mathematics together with the principles of differential and integral calculus. In doing so I had the good fortune of finding books which were not too particular in their logical rigor, but which made up for this by permitting the main thoughts to stand out clearly and synoptically . . . I also had the good fortune of getting to know the essential results and methods of the entire field of the natural sciences in an excellent popular exposition, which limited itself almost throughout to qualitative aspects . . . a work which I read with breathless attention." Modern popularizers of science may take some comfort from these words.

Not one of his teachers seems to have recognized his talents. At the Munich *Gymnasium,* the city's leading secondary school, one of the teachers told him, "You'll never amount to anything, Einstein." At age fifteen it was strongly suggested that he leave school. The teacher observed, "Your very presence spoils the respect of the class for me." He accepted this suggestion with gusto and spent many months wandering through northern Italy, a high school dropout in the 1890s. Throughout his life he preferred informal dress and manner. Had he been a teen-ager in the 1960s or 1970s rather than the 1890s, conventional people would almost certainly have called him a hippie.

Yet his curiosity about physics and his wonder about the natural universe soon overcame his distaste for formal education, and he found himself applying, with no high school diploma, to the Federal Institute of Technology in Zurich, Switzerland. He failed the entrance examination, enrolled himself in a Swiss high school to satisfy his deficiencies, and was admitted to the Federal Institute the following year. But he was still a mediocre student. He resented the prescribed curriculum, avoided the lecture room and tried to pursue his true interests. He later wrote: "The hitch in this was, of course, the fact that you had to cram all this stuff into your mind for the examination, whether you liked it or not."

He managed to graduate only because his close friend Marcel Grossmann assiduously attended classes and shared his notes with Einstein. On Grossmann's death many years later, Einstein wrote: "I remember our student days. He the irreproachable student, I myself disorderly and a dreamer. He, on good terms with the teachers and understanding everything; I a pariah, discontented and little loved . . . Then the end of our studies—I was suddenly abandoned by everyone, standing at a loss on the threshold of life." By immersing himself in Grossmann's notes, he managed to graduate from college. But, he recalled, studying for the final examinations "had such a deterring effect on me that . . . I found the consideration of any scientific problem distasteful to me for an entire year. . . . It is little short of a miracle that modern methods of instruction have not already completely strangled the holy curiosity of inquiry, because what this delicate little plant needs most, apart from initial stimulation, is freedom; without that it is surely destroyed . . . I believe that one could even deprive a healthy beast of prey of its voraciousness, if one could force it with a whip to eat continuously whether it were hungry or not . . ." His remarks should be sobering to those of us engaged in higher education in science. I wonder how many potential Einsteins have been permanently discouraged through competitive examinations and the forced feeding of curricula.

After supporting himself with odd jobs, and being passed over for positions he considered desirable, Einstein accepted an offer as an examiner of applications at the Swiss Patent Office in Berne, an opportunity made available through the intervention of Marcel Grossmann's father. About the same time he rejected his German nationality and became a Swiss citizen. Three years later, in 1903, he married his college sweetheart. Almost nothing is known about which patent applications Einstein approved and which he rejected. It would be interesting to know whether any of the proposed patents stimulated his thinking in physics.

One of his biographers, Banesh Hoffman, writes that at the Patent Office, Einstein "soon learned to do his chores efficiently and this let him snatch precious morsels of time for his own surreptitious calculations, which he guiltily hid in a drawer when footsteps approached." Such were the circumstances attending the birth of the great relativity theory. But Einstein later nostalgically recalled the Patent Office as "that secular cloister where I hatched my most beautiful ideas."

On several occasions he was to suggest to colleagues that the occupation of lighthouse keeper would be a suitable position for a scientist—because the work would be comparatively easy and would allow the contemplation necessary to do scientific research. "For Einstein," said his collaborator Leopold Infeld, "loneliness, life in a lighthouse, would be most stimulating, would free him from so many of the duties which he hates. In fact it would be for him the ideal life. But nearly every scientist thinks just the opposite. It was the curse of *my* life that for a long time I was not in a scientific atmosphere, that I had no one with whom to talk physics."

Einstein also believed that there was something dishonest about making money by teaching physics. He argued that it was far better for a physicist to support himself by some other simple and honest labor, and do physics in his spare time. When making a similar remark many years later in America, Einstein mused that he would have liked to be a plumber, and was promptly awarded honorary membership in the plumbers' union.

In 1905 Einstein published four research papers, the product of his spare time at the Swiss Patent Office, in the leading physics journal of the time, the *Annalen der Physik*. The first demonstrated that light has particle as well as wave properties, and explained the previously baffling photoelectric effect in which electrons are emitted by solids when irradiated by light. The second explored the nature of molecules by explaining the statistical "Brownian motion" of suspended small particles. And the third and fourth introduced the Special Theory of Relativity and for the first time expressed the famous equation, $E = mc^2$, which is so widely quoted and so rarely understood.

The equation expresses the convertibility of matter into energy, and vice versa. It extends the law of the conservation of energy into a law of conservation of energy and mass, stating that energy and mass can be neither created nor destroyed—although one form of energy or matter can be converted into another form. In the equation, E stands for the energy equivalent of the mass m. The amount of energy that could, under ideal circumstances, be extracted from a mass m is mc^2, where c is the velocity of light = 30 billion centimeters per second. (The velocity of light is always written as lower-case c, never as upper-case.) If we measure m in grams and c in centimeters per second, E is measured in a unit of energy called ergs. The complete conversion of one gram of mass into energy thus releases $1 \times (3 \times 10^{10})^2 = 9 \times 10^{20}$ ergs, which is the equivalent of the explosion of roughly a thousand tons of TNT. Thus enormous energy resources are contained in tiny amounts of matter, if only we knew how to extract the energy. Nuclear weapons and nuclear power plants are common terrestrial examples of our halting and ethically ambiguous efforts to extract the energy that Einstein showed was present in all of matter. A thermonuclear weapon, a hydrogen bomb, is a device of terrifying power—but even it is capable of extracting less than one percent of mc^2 from a mass m of hydrogen.

Einstein's four papers published in 1905 would have been an impressive output for the full-time research work of a physicist over a lifetime; for the spare-time work of a twenty-six-year-old Swiss patent clerk in a single year it is nothing short of astonishing. Many historians of science have called 1905 the *Annus Mirabilis*, the miracle year. There had been, with uncanny resemblances, only one previous such year in the history of physics—1666, when Isaac Newton, aged twenty-four, in enforced rural isolation (because of an epidemic of bubonic plague) produced an explanation for the spectral nature of sunlight, invented differential and integral calculus, and devised the universal theory of gravitation. Together with the General Theory of Relativity, first formulated in 1915, the 1905 papers represent the principal output of Einstein's scientific life.

Before Einstein, it was widely held by physicists that there were privileged frames of reference, such things as absolute space and absolute time. Einstein's starting point was that all frames of reference—all observers, no matter what their locale, velocity or acceleration—would see the fundamental laws of nature in the same way. It seems likely that Einstein's view on frames of reference was influenced by his social and political attitudes and his resistance to the strident jingoism he found in late-nineteenth-century Germany. Indeed, in this sense the idea of relativity has become an anthropological commonplace, and social scientists have adopted the idea of cultural relativism: there are many different social contexts and world views, ethical and religious precepts, expressed by various human societies, and most of comparable validity.

Special relativity was at first by no means widely accepted. Attempting once again to break into an academic career, Einstein submitted his already published relativity paper to Berne University as an example of his work. He evidently considered it a significant piece of research. It was rejected as incomprehensible, and he was to remain at the Patent Office until 1909. But his published work did not go unnoticed, and it slowly began to dawn on a few of the leading European physicists that Einstein might well be one of the greatest scientists of all time. Still, his work on relativity remained highly controversial. In a letter of recommendation for Einstein for a position at the University of Berlin, a leading German scientist suggested that relativity was a hypothetical excursion, a momentary aberration, and that, despite it, Einstein really *was* a first-rate thinker. (His Nobel Prize, which he learned about during a visit to the Orient in 1921, was awarded for his paper on the photoelectric effect and "other contributions" to theoretical physics. Relativity was still considered too controversial to be mentioned explicitly.)

Einstein's views on religion and politics were connected. His parents were of Jewish origin, but they did not observe religious ritual. Nevertheless, Einstein came to a conventional religiosity "by way of the traditional education machine, the State and the schools." But at age twelve this came to an abrupt end: "Through the reading of popular scientific books I soon reached the conviction that much of the stories of the Bible could not be true. The consequence was a positively fanatic free thinking coupled with the impression that youth is intentionally being deceived by the State through lies; it was a crushing impression. Suspicion against every kind of authority grew out of this experience, a skeptical attitude towards the convictions which were alive in any specific social environment—an attitude which has never again left me, even though later on, because of a better insight into the causal connections, it lost some of its original poignancy."

Just before the outbreak of World War I, Einstein accepted a professorship at the well-known Kaiser Wilhelm Institute in Berlin. The desire to be at the leading center of theoretical physics was momentarily stronger than his antipathy to German militarism. The outbreak of World War I caught Einstein's wife and two sons in Switzerland, unable to return to Germany. A few years later this enforced separation led to divorce, but on receiving the Nobel Prize in 1921, Einstein, although since remarried, donated the full $30,000 to his first wife and their children. His eldest son later became a significant figure in civil engineering, holding a professorship at the University of California, but his second son, who idolized his father, accused him—in later years, and to Einstein's great anguish—of having ignored him during his youth.

Einstein, who described himself as a socialist, became convinced that World War I was largely the result of the scheming and incompetence of "the ruling classes," a conclusion with which many contemporary historians agree. He became a pacifist. When other German scientists enthusiastically supported their nation's military enterprises, Einstein publicly condemned the war as "an epidemic delusion." Only his Swiss citizenship prevented him from being imprisoned, as indeed happened to his friend the philosopher Bertrand Russell in England at the same time and for the same reason. Einstein's views on the war did not increase his popularity in Germany.

However, the war did, indirectly, play a role in making Einstein's name a household word. In his General Theory of Relativity Einstein explored the proposition—an idea still astonishing in its simplicity, beauty and power—that the gravitational attraction between two masses comes about by those masses distorting or bending ordinary Euclidean space nearby. The quantitative theory reproduced, to the accuracy to which it had been tested, Newton's law of universal gravitation. But in the next decimal place, so to speak, general relativity predicted significant differences from Newton's views. This is in the classic tradition of science, in which new theories retain the established results of the old but make a set of new predictions which permits a decisive distinction to be drawn between the two outlooks.

The three tests of general relativity that Einstein proposed concerned anomalies in the motion of the orbit of the planet Mercury, the red shifts in the spectral lines of light emitted by a massive star, and the deflection of starlight near the Sun. Before the Armistice was signed in 1919, British expeditions were mustered to Brazil and to the island of Principe off West Africa to observe, during a total eclipse of the Sun, whether the deflection of starlight was in accord with the predictions of general relativity. It was. Einstein's views were vindicated; and the symbolism of a British expedition confirming the work of a German scientist when the two countries were still technically at war appealed to the better instincts of the public.

But at the same time, a well-financed public campaign against Einstein was launched in Germany. Mass meetings with anti-Semitic overtones were staged in Berlin and elsewhere to denounce the relativity theory. Einstein's colleagues were shocked, but most of them, too timid for politics, did nothing to counter it. With the rise of the Nazis in the 1920s and early 1930s, Einstein, against his natural inclination for a life of quiet contemplation, found himself speaking up—courageously and often. He testified in German courts on behalf of academics on trial for their political views. He appealed for amnesty for political prisoners in Germany and abroad (including Sacco and Vanzetti and the Scottsboro "boys" in the United States). When Hitler became chancellor in 1933, Einstein and his second wife fled Germany.

The Nazis burned Einstein's scientific works, along with other books by anti-Fascist authors, in public bonfires. An all-out assault was launched on Einstein's scientific stature. Leading the attack was the Nobel laureate physicist Philipp Lenard, who denounced what he called the "mathematically botched-up theories of Einstein" and the "Asiatic spirit in Science." He went on: "Our Führer has eliminated this same spirit in politics and national economy, where it is known as Marxism. In natural science, however, with the overemphasis on Einstein, it still holds sway. We must recognize that it is unworthy of a German to be the intellectual follower of a Jew. Natural science, properly so-called, is of completely aryan origin . . . *Heil Hitler!*"

Many Nazi scholars joined in warning against the "Jewish" and "Bolshevik" physics of Einstein. Ironically, in the Soviet Union at about the same time, prominent Stalinist intellectuals were denouncing relativity as "bourgeois physics." Whether or not the *substance* of the theory being attacked was correct was, of course, never considered in such deliberations.

Einstein's identification of himself as a Jew, despite his profound estrangement from traditional religions, was due entirely to the upsurge of anti-Semitism in Germany in the 1920s. For this reason he also became a Zionist. But according to his biographer Philipp

Frank, not all Zionist groups welcomed him, because he demanded that the Jews make an effort to befriend the Arabs and to understand their way of life—a devotion to cultural relativism made more impressive by the difficult emotional issues involved. However, he continued to support Zionism, particularly as the increasing desperation of European Jews became known in the late 1930s. (In 1948 Einstein was offered the presidency of Israel, but politely declined. It is interesting to speculate what differences in the politics of the Near East, if any, might have been produced by Albert Einstein as the president of Israel.)

After leaving Germany, Einstein learned that the Nazis had placed a price of 20,000 marks on his head. ("I didn't know it was worth so much.") He accepted an appointment at the recently founded Institute for Advanced Study in Princeton, New Jersey, where he was to remain for the rest of his life. When asked what salary he thought fair, he suggested $3,000. Seeing a look of astonishment pass over the face of the representative of the Institute, he concluded he had proposed too much and mentioned a smaller amount. His salary was set at $16,000, a goodly sum for the 1930s.

Einstein's prestige was so high that it was natural for other émigré European physicists in the United States to approach him in 1939 to write a letter to President Franklin D. Roosevelt, proposing the development of an atomic bomb to outstrip a likely German effort to acquire nuclear weapons. Although Einstein had not been working in nuclear physics and later played no role in the Manhattan Project, he wrote the initial letter that led to the establishment of the Manhattan Project. It is likely, however, that the bomb would have been developed by the United States regardless of Einstein's urging. Even without $E = mc^2$, the discovery of radioactivity by Antoine Becquerel and the investigation of the atomic nucleus by Ernest Rutherford—both done entirely independently of Einstein—would very likely have led to the development of nuclear weapons. Einstein's dread of Nazi Germany had long since caused him to abandon, although with considerable pain, his pacifist views. But when it later transpired that the Nazis had been unable to develop nuclear weapons, Einstein expressed remorse: "Had I known that the Germans would not succeed in developing an atomic bomb, I would have done nothing for the bomb."

In 1945 Einstein urged the United States to break its relations with Franco Spain, which had supported the Nazis in World War II. John Rankin, a conservative congressman from Mississippi, attacked Einstein in a speech to the House of Representatives, declaring that "this foreign-born agitator would have us plunge into another war in order to further the spread of Communism throughout the world . . . It is about time the American people got wise to Einstein."

Einstein was a powerful defender of civil liberties in the United States during the darkest period of McCarthyism in the late 1940s and early 1950s. Watching the rising tide of hysteria, he had the disturbing feeling that he had seen something similar in Germany in the 1930s. He urged defendants to refuse to testify before the House Un-American Activities Committee, saying that every person should be "prepared for jail and economic ruin . . . for the sacrifice of his personal welfare in the interest of . . . his country." He held that there was "a duty in refusing to cooperate in any undertaking that violates the Constitutional rights of the individual. This holds in particular for all inquisitions that are concerned with the private life and the political affiliations of the citizens . . ." For taking this position, Einstein was widely attacked in the press. And Senator Joseph McCarthy stated in 1953 that anyone who proffered such advice was "himself an enemy of America." In his later years it became fashionable in some circles to couple an acknowledgment of Einstein's scientific genius with a patronizing dismissal of his political views as "naïve." But times have changed. I wonder if it is not more reasonable to argue in quite a different direction: in a field such as physics, where ideas can be quantified and tested with great precision, Einstein's insights stand unrivaled, and we are astonished that he could see so clearly where others were lost in confusion. Is it not worth considering that in the much murkier field of politics his insights might also have some fundamental validity?

In his Princeton years Einstein's passion remained, as always, the life of the mind. He worked long and hard on a Unified Field Theory which would combine gravitation, electricity and magnetism on a common basis, but his attempt is widely considered to have been unsuccessful. He lived to see his General Theory of Relativity incorporated as the principal tool for understanding the large-scale structure and evolution of the universe, and would have been delighted to witness the vigorous application of general relativity occurring in astrophysics today. He never understood the reverence with which he was held, and indeed complained that his colleagues and Princeton graduate students would not drop in on him unannounced for fear of disturbing him.

But he wrote: "My passionate interest in social justice and social responsibility has always stood in curious contrast to a marked lack of desire for direct association with men and women. I am a horse for single harness, not cut out for tandem or team work. I have never belonged wholeheartedly to country or State, to my circle of friends or even to my own family. These ties have always been accompanied by a vague aloofness, and the wish to withdraw into myself increases with the years. Such isolation is sometimes bitter, but I do not regret being cut off from the understanding and sympathy of other men. I lose something by it, to be sure, but I am compensated for it in being rendered independent of the customs, opinions and prejudices of others and am not tempted to rest my peace of mind upon such shifting foundations."

His principal recreations throughout his life were playing the violin and sailing. In these years Einstein looked like and in some respects was a sort of aging hippie. He let his white hair grow long and preferred sweaters and a leather jacket to a suit and tie, even when entertaining famous visitors. He was utterly without pretense and, with no affectation, explained that "I speak to everyone in the same way, whether he is the garbage man or the President of the University." He was often available to the public, sometimes being willing to help high school students with their geometry problems—not always successfully. In the best scientific tradition he was open to new ideas but required that they pass rigorous standards of evidence. He was open-minded but skeptical about claims of planetary catastrophism in recent Earth history and about experiments alleging extrasensory perception, his reservations about the latter stemming from contentions that purported telepathic abilities do not decline with increasing distance between sender and receiver.

In matters of religion, Einstein thought more deeply than many others and was repeatedly misunderstood. On the occasion of Einstein's first visit to America, Cardinal O'Connell of Boston warned that the relativity theory "cloaked the ghastly apparition of atheism." This alarmed a New York rabbi who cabled Einstein: "Do you believe in God?" Einstein cabled back: "I believe in Spinoza's God, who revealed himself in the harmony of all being, not in the God who concerns himself with the fate and actions of men"—a more subtle religious view embraced by many theologians today. Einstein's religious beliefs were very genuine. In the 1920s and 1930s he expressed grave doubts about a basic precept of quantum mechanics: that at the most fundamental level of matter, particles behave in an unpredictable way, as expressed by the Heisenberg uncertainty principle. Einstein said, "God does not play dice with the cosmos." And on another occasion he asserted, "God is subtle, but he is not malicious." In fact, Einstein was so fond of such aphorisms that the Danish physicist Niels Bohr turned to him on one occasion and with some exasperation said, "Stop telling God what to do." But there were many physicists who felt that if anyone knew God's intentions, it was Einstein.

One of the foundations of special relativity is the precept that no material object can travel as fast as light.

This light barrier has proved annoying to many people who wish there to be no constraints on what human beings might ultimately do. But the light limit permits us to understand much of the world that was previously mysterious in a simple and elegant way. However, where Einstein taketh away, he also giveth. There are several consequences of special relativity that seem counterintuitive, contrary to our everyday experience, but that emerge in a detectable fashion when we travel close to the speed of light—a regime of velocity in which common sense has had little experience (Chapter 2). One of these consequences is that as we travel sufficiently close to the speed of light, time slows down—our wristwatches, our atomic clocks, our biological aging. Thus a space vehicle traveling very close to the speed of light could travel between any two places, no matter how distant, in any conveniently short period of time—as measured on board the spacecraft, but not as measured on the launch planets. We might therefore one day travel to the center of the Milky Way Galaxy and return in a time of a few decades measured on board the ship—although, as measured back on Earth, the elapsed time would be sixty thousand years, and very few of the friends who saw us off would be around to commemorate our return. A vague recognition of this time dilation was made in the motion picture *Close Encounters of the Third Kind*, although a gratuitous opinion was then injected that Einstein was probably an extraterrestrial. His insights were stunning, to be sure, but he was very human, and his life stands as an example of what, if they are sufficiently talented and courageous, human beings can accomplish.

EINSTEIN'S LAST public act was to join with Bertrand Russell and many other scientists and scholars in an unsuccessful attempt to bring about a ban on the development of nuclear weapons. He argued that nuclear weapons had changed everything except our way of thinking. In a world divided into hostile states he viewed nuclear energy as the greatest menace to the survival of the human race. "We have the choice," he said, "to outlaw nuclear weapons or face general annihilation. . . . Nationalism is an infantile disease. It is the measles of mankind . . . Our schoolbooks glorify war and hide its horrors. They inculcate hatred in the veins of children. I would teach peace rather than war. I would inculcate love rather than hate."

At age sixty-seven, nine years before his death in 1955, Einstein described his lifelong quest: "Out yonder there was this huge world, which exists independently of us human beings and which stands before us like a great, eternal riddle, at least partially accessible to our inspection and thinking. The contemplation of this world beckoned like a liberation . . . The road to this paradise was not so comfortable and alluring as the road to the religious paradise; but it has proved itself as trustworthy, and I have never regretted having chosen it."

CHAPTER 4

IN PRAISE OF SCIENCE AND TECHNOLOGY

◆◆◆

The cultivation of the mind is a kind of food supplied for the soul of man.

MARCUS TULLIUS CICERO,
De Finibus Bonorum et Malorum,
Vol. 19 (45–44 B.C.)

To one, science is an exalted goddess; to another it is a cow which provides him with butter.

FRIEDRICH VON SCHILLER,
Xenien (1796)

IN THE MIDDLE of the nineteenth century, the largely self-educated British physicist Michael Faraday was visited by his monarch, Queen Victoria. Among Faraday's many celebrated discoveries, some of obvious and immediate practical benefit, were more arcane findings in electricity and magnetism, then little more than laboratory curiosities. In the traditional dialogue between heads of state and heads of laboratories, the Queen asked Faraday of what use such studies were, to which he is said to have replied, "Madam, of what use is a baby?" Faraday had an idea that there might someday be something practical in electricity and magnetism.

In the same period the Scottish physicist James Clerk Maxwell set down four mathematical equations, based on the work of Faraday and his experimental predecessors, relating electrical charges and currents with electric and magnetic fields. The equations exhibited a curious lack of symmetry, and this bothered Maxwell. There was something unaesthetic about the equations as then known, and to improve the symmetry Maxwell proposed that one of the equations should have an additional term, which he called the displacement current. His argument was fundamentally intuitive; there was certainly no experimental evidence for such a current. Maxwell's proposal had astonishing consequences. The corrected Maxwell equations implied the existence of electromagnetic radiation, encompassing gamma rays, X-rays, ultraviolet light, visible light, infrared and radio. They stimulated Einstein to discover Special Relativity. Faraday and Maxwell's laboratory and theoretical work together have led, one century later, to a technical revolution on the planet Earth. Electric lights, telephones, phonographs, radio, television, refrigerated trains making fresh produce available far from the farm, cardiac pacemakers, hydroelectric power plants, automatic fire alarms and sprinkler systems, electric trolleys and subways, and the electronic computer are a few devices in the direct evolutionary line from the arcane laboratory puttering of Faraday and the aesthetic dissatisfaction of Maxwell, staring at some mathematical squiggles on a piece of paper. Many of the most practical applications of science have been made in this serendipitous and unpredictable way. No amount of money would have sufficed in Victoria's day for the leading scientists in Britain to have simply sat down and invented, let us say, television. Few would argue that the net effect of these inventions was other than positive. I notice that even many young people who are profoundly disenchanted with Western technological civilization, often for good reason, still retain a passionate fondness for certain aspects of high technology—for example, high-fidelity electronic music systems.

Some of these inventions have fundamentally changed the character of our global society. Ease of communication has deprovincialized many parts of the world, but cultural diversity has been likewise diminished. The practical advantages of these inventions are recognized in virtually all human societies; it is remarkable how infrequently emerging nations are concerned with the negative effects of high technology (environmental pollution, for example); they have clearly decided that the benefits outweigh the risks. One of Lenin's aphorisms was that socialism plus electrification equals communism. But there has been no more vigorous or inventive pursuit of high technology than in the West. The resulting rate of change has been so rapid that many of us find it difficult to keep up. There are many people alive today who were born before the first airplane and have lived to see Viking land on Mars, and Pioneer 10, the first interstellar spacecraft, be ejected from the solar system, or who were raised in a sexual code of Victorian severity and now find themselves immersed in substantial sexual freedom, brought about by the widespread availability of effective contraceptives. The rate of change has been disorienting for many, and it is easy to understand the nostalgic appeal of a return to an earlier and simpler existence.

But the standard of living and conditions of work for the great bulk of the population in, say, Victorian England, were degrading and demoralizing compared to industrial societies today, and the life-expectancy and infant-mortality statistics were appalling. Science and technology may be in part responsible for many of the problems that face us today—but largely because public understanding of them is desperately inadequate (technology is a tool, not a panacea), and because insufficient effort has been made to accommodate our society to the new technologies. Considering these facts, I find it remarkable that we have done as well as we have. Luddite alternatives can solve nothing. More than one billion people alive today owe the margin between barely adequate nutrition and starvation to high agricultural technology. Probably an equal number have survived, or avoided disfiguring, crippling or killing diseases because of high medical technology. Were high technology to be abandoned, these people would also be abandoned. Science and technology may be the cause of some of our problems, but they are certainly an essential element in any foreseeable solution to those same problems—both nationally and planetwide.

I do not think that science and technology have been pursued as effectively, with as much attention to their ultimate humane objectives and with as adequate a public understanding as, with a little greater effort, could have been accomplished. It has, for example, gradually dawned on us that human activities can have an adverse effect on not only the local but also the global environment. By accident a few research groups in atmospheric photochemistry discovered that halocarbon propellants from aerosol spray cans will reside for very long periods in the atmosphere, circulate to the stratosphere, partially destroy the ozone there, and let ultraviolet light from the sun leak down to the Earth's surface. Increased skin cancer for whites was the most widely advertised consequence (blacks are neatly adapted to increased ultraviolet flux). But very little public attention has been given to the much more serious possibility that microorganisms, occupying the base of an elaborate food pyramid at the top of which is *Homo sapiens,* might also be destroyed by the increased ultraviolet light. Steps have finally, although reluctantly, been taken to ban halocarbons from spray cans (although no one seems to be worrying about the same molecules used in refrigerators) and as a result the immediate dangers are probably slight. What I find most worrisome about this incident is how accidental was the discovery that the problem existed at all. One group approached this problem because it had written the appropriate computer programs, but in quite a different context: they were concerned with the chemistry of the atmosphere of the planet Venus, which contains hydrochloric and hydrofluoric acids. The need for a broad and diverse set of research teams, working on a great variety of problems in pure science, is clearly required for our continued survival. But what other problems, even more severe, exist which we do not know about because no research group happens as yet to have stumbled on them? For each problem we have uncovered, such as the effect of halocarbons on the ozonosphere, might there not be another dozen lurking around the corner? It is therefore an astonishing fact that nowhere in the federal government, major universities or private research institutes is there a single highly competent, broadly empowered and adequately funded research group whose function it is to seek out and defuse future catastrophes resulting from the development of new technologies.

The establishment of such research and environmental assessment organizations will require substantial political courage if they are to be effective at all. Technological societies have a tightly knit industrial ecology, an interwoven network of economic assumptions. It is very difficult to challenge one thread in the network without causing tremors in all. Any judgment that a technological development will have adverse human consequences implies a loss of profit for someone. The DuPont Company, the principal manufacturers of halocarbon propellants, for example, took the curious position in public debates that all conclusions about halocarbons destroying the ozonosphere were "theoretical." They seemed to be implying that they would be prepared to stop halocarbon manufacture only after the conclusions were tested experimentally—that is, when the ozonosphere was destroyed. There are some problems where inferential evidence is all that we will have; where once the catastrophe arrives it is too late to deal with it.

Similarly, the new Department of Energy can be effective only if it can maintain a distance from vested commercial interests, if it is free to pursue new options even if such options imply loss of profits for selected industries. The same is clearly true in pharmaceutical research, in the pursuit of alternatives to the internal-combustion engine, and in many other technological frontiers. I do not think that the development of new technologies should be placed in the control of old technologies; the temptation to suppress the competition is too great. If we Americans live in a free-enterprise society, let us see substantial independent enterprise in all of the technologies upon which our future may depend. If organizations devoted to technological innovation and its boundaries of acceptability are not challenging (and perhaps even offending) at least *some* powerful groups, they are not accomplishing their purpose.

There are many practical technological developments that are not being pursued for lack of government support. For example, as agonizing a disease as cancer is, I do not think it can be said that our civilization is threatened by it. Were cancer to be cured completely, the average life expectancy would be extended by only a few years, until some other disease—which does not now have its chance at cancer victims—takes over. But a very plausible case can be made that our civilization is fundamentally threatened by the lack of adequate fertility control. Exponential increases of population will dominate any arithmetic increases, even those brought about by heroic technological initiatives, in the availability of food and resources, as Malthus long ago realized. While some industrial nations have approached zero population growth, this is not the case for the world as a whole.

Minor climatic fluctuations can destroy entire populations with marginal economies. In many societies where the technology is meager and reaching adulthood an uncertain prospect, having many children is the only possible hedge against a desperate and uncertain future. Such a society, in the grip of a consuming famine, for example, has little to lose. At a time when nuclear weapons are proliferating unconscionably, when an atomic device is almost a home handicraft industry, widespread famine and steep gradients in affluence pose serious dangers to both the developed and the underdeveloped worlds. The solution to such problems certainly requires better education, at least a degree of technological self-sufficiency, and, especially, fair distribution of the world's resources. But it also cries out for entirely adequate contraception—long-term, safe birth-control pills, available for men as well as for women, perhaps to be taken once a month or over even longer intervals. Such a development would be very useful not just abroad but also here at home, where considerable concern is being expressed about the side effects of the conventional estrogen oral contraceptives. Why is there no major effort for such a development?

Many other technological initiatives are being proposed and ought to be examined very seriously. They range from the very cheap to the extremely expensive. At one end is soft technology—for example, the development of closed ecological systems involving algae, shrimp and fish which could be maintained in rural ponds and provide a highly nutritious and extremely low-cost dietary supplement. At the other is the proposal of Gerard O'Neill of Princeton University to construct large orbital cities that would, using lunar and asteroidal materials, be self-propagating—one city being able to construct another from extraterrestrial resources. Such cities in Earth orbit might be used in converting sunlight into microwave energy and beaming power down to Earth. The idea of independent cities in space—each perhaps built on differing social, economic or political assumptions, or having different ethnic antecedents—is appealing, an opportunity for those deeply disenchanted with terrestrial civilizations to strike out on their own somewhere else. In its earlier history, America provided such an opportunity for the restless, ambitious and adventurous. Space cities would be a kind of America in the skies. They also would greatly enhance the survival potential of the human species. But the project is extremely expensive, costing at minimum about the same as one Vietnam war (in resources, not in lives). In addition, the idea has the worrisome overtone of abandoning the problems on the Earth—

where, after all, self-contained pioneering communities can be established at much less cost.

Clearly, there are more technological projects now possible than we can afford. Some of them may be extremely cost-effective but may have such large start-up costs as to remain impractical. Others may require a daring initial investment of resources, which will work a benevolent revolution in our society. Such options have to be considered extremely carefully. The most prudent strategy calls for combining low-risk/moderate-yield and moderate-risk/high-yield endeavors.

For such technological initiatives to be understood and supported, significant improvements in public understanding of science and technology are essential. We are thinking beings. Our minds are our distinguishing characteristic as a species. We are not stronger or swifter than many other animals that share this planet with us. We are only smarter. In addition to the immense practical benefit of having a scientifically literate public, the contemplation of science and technology permits us to exercise our intellectual faculties to the limits of our capabilities. Science is an exploration of the intricate, subtle and awesome universe we inhabit. Those who practice it know, at least on occasion, a rare kind of exhilaration that Socrates said was the greatest of human pleasures. It is a communicable pleasure. To facilitate informed public participation in technological decision making, to decrease the alienation too many citizens feel from our technological society, and for the sheer joy that comes from knowing a deep thing well, we need better science education, a superior communication of its powers and delights. A simple place to start is to undo the self-destructive decline in federal scholarships and fellowships for science researchers and science teachers at the college, graduate and postdoctoral levels.

The most effective agents to communicate science to the public are television, motion pictures and newspapers—where the science offerings are often dreary, inaccurate, ponderous, grossly caricatured or (as with much Saturday-morning commercial television programing for children) hostile to science. There have been astonishing recent findings on the exploration of the planets, the role of small brain proteins in affecting our emotional lives, the collisions of continents, the evolution of the human species (and the extent to which our past prefigures our future), the ultimate structure of matter (and the question of whether there are elementary particles or an infinite regress of them), the attempt to communicate with civilizations on planets of other stars, the nature of the genetic code (which determines our heredity and makes us cousins to all the other plants and animals on our planet), and the ultimate questions of the origin, nature and fate of life, worlds and the universe as a whole. Recent findings on these questions can be understood by any intelligent person. Why are they so rarely discussed in the media, in schools, in everyday conversation?

Civilizations can be characterized by how they approach such questions, how they nourish the mind as well as the body. The modern scientific pursuit of these questions represents an attempt to acquire a generally accepted view of our place in the cosmos; it requires open-minded creativity, tough-minded skepticism and a fresh sense of wonder. These questions are different from the practical issues I discussed earlier, but they are connected with such issues and—as in the example of Faraday and Maxwell—the encouragement of pure research may be the most reliable guarantee available that we will have the intellectual and technical wherewithal to deal with the practical problems facing us.

Only a small fraction of the most able youngsters enter scientific careers. I am often amazed at how much more capability and enthusiasm for science there is among elementary school youngsters than among college students. Something happens in the school years to discourage their interest (and it is not mainly puberty); we must understand and circumvent this dangerous discouragement. No one can predict where the future leaders of science will come from. It is clear that Albert Einstein became a scientist in spite of, not because of, his schooling (Chapter 3). In his *Autobiography*, Malcolm X describes a numbers runner who never wrote down a bet but carried a lifetime of transactions perfectly in his head. What contributions to society, Malcolm asked, would such a person have made with adequate education and encouragement? The most brilliant youngsters are a national and a global resource. They require special care and feeding.

Many of the problems facing us may be soluble, but only if we are willing to embrace brilliant, daring and complex solutions. Such solutions require brilliant, daring and complex people. I believe that there are many more of them around—in every nation, ethnic group and degree of affluence—than we realize. The training of such youngsters must not, of course, be restricted to science and technology; indeed, the compassionate application of new technology to human problems requires a deep understanding of human nature and human culture, a general education in the broadest sense.

We are at a crossroads in human history. Never before has there been a moment so simultaneously perilous and promising. We are the first species to have taken our evolution into our own hands. For the first time we possess the means for intentional or inadvertent self-destruction. We also have, I believe, the means for passing through this stage of technological adolescence into a long-lived, rich and fulfilling maturity for all the members of our species. But there is not much time to determine to which fork of the road we are committing our children and our future.

PART II

THE PARADOXERS

CHAPTER 5

NIGHT WALKERS AND MYSTERY MONGERS: SENSE AND NONSENSE AT THE EDGE OF SCIENCE

◆◆◆

PLANT'S HEARTBEAT THRILLS SCIENTISTS AT OXFORD MEETING
Hindu Savant causes further sensation by showing "blood" of plant flowing

AUDIENCE SITS ABSORBED
Watches with rapt attention as lecturer submits snapdragon to death struggle

The New York Times
August 7, 1926, page 1

William James used to preach the
"will to believe."
For my part, I should wish
to preach the "will to doubt." . . .
What is wanted is not the will to believe,
but the wish to find out, which is
the exact opposite.

BERTRAND RUSSELL,
Sceptical Essays (1928)

IN GREECE of the second century A.D., during the reign of the Roman Emperor Marcus Aurelius, there lived a master con man named Alexander of Abonutichus. Handsome, clever and totally unscrupulous, in the words of one of his contemporaries, he "went about living on occult pretensions." In his most famous imposture, "he rushed into the marketplace, naked except for a gold-spangled loincloth; with nothing but this and his scimitar, and shaking his long, loose hair, like fanatics who collect money in the name of Cybele, he climbed onto a lofty altar and delivered a harangue" predicting the advent of a new and oracular god. Alexander then raced to the construction site of a temple, the crowd streaming after him, and discovered—where he had previously buried it—a goose egg in which he had sealed up a baby snake. Opening the egg, he announced the snakelet as the prophesied god. Alexander retired to his house for a few days, and then admitted the breathless crowds, who observed his body now entwined with a large serpent: the snake had grown impressively in the interim.

The serpent was, in fact, of a large and conveniently docile variety, procured for this purpose earlier in Macedonia, and outfitted with a linen head of somewhat human countenance. The room was dimly lit. Because of the press of the crowd, no visitor could stay for very long or inspect the serpent very carefully. The opinion of the multitude was that the seer had indeed delivered a god.

Alexander then pronounced the god ready to answer written questions delivered in sealed envelopes. When alone, he would lift off or duplicate the seal, read the message, remake the envelope and attach a response. People flocked from all over the Empire to witness this marvel, an oracular serpent with the head of a man. In those cases where the oracle later proved not just ambiguous but grossly wrong, Alexander had a simple solution: he altered his record of the response he had given. And if the question of a rich man or woman revealed some weakness or guilty secret, Alexander did not scruple at extortion. The result of all this imposture was an income equivalent today to several hundred thousand dollars per year and fame rivaled by few men of his time.

We may smile at Alexander the Oracle-Monger. Of course we all would like to foretell the future and make contact with the gods. But we would not nowadays be taken in by such a fraud. Or would we? M. Lamar Keene spent thirteen years as a spiritualist medium. He was pastor of the New Age Assembly Church in Tampa, a trustee of the Universal Spiritualist Association, and for many years a leading figure in the mainstream of the American spiritualist movement. He is also a self-confessed fraud who believes, from first-hand knowledge, that virtually all spirit readings, séances and mediumistic messages from the dead are conscious deceptions, contrived to exploit the grief and longing we feel for deceased friends and relatives. Keene, like Alexander, would answer questions given to him in sealed envelopes—in this case not in private, but on the pulpit. He viewed the contents with a concealed bright lamp or by smearing lighter fluid, either of which can render the envelope momentarily transparent. He would find lost objects, present people with astounding revelations about their private lives which "no one could know," commune with the spirits and materialize ectoplasm in the darkness of the séance—all based on the simplest tricks, an unswerving self-confidence, and most of all, on the monumental credulity, the utter lack of skepticism he found in his parishioners and clients. Keene believes, as did Harry Houdini, that not only is such fraud rampant among the spiritualists, but also that they are highly organized to exchange data on potential clients, in order to make the revelations of the séance more astonishing. Like the viewing of Alexander's serpent, the séances all take place in darkened rooms—because the deception would be too easily penetrated in the light. In his peak-earning years, Keene earned about as much, in equivalent purchasing power, as Alexander of Abonutichus.

From Alexander's time to our own—indeed, probably for as long as human beings have inhabited this planet—people have discovered they could make money by pretending to arcane or occult knowledge. A charming and enlightening account of some of these bamboozles can be found in a remarkable book published in 1852 in London, *Extraordinary Popular Delusions and the Madness of Crowds,* by Charles Mackay. Bernard Baruch claimed that the book saved him millions of dollars—presumably by alerting him to which idiot schemes he should not invest his money in. Mackay's treatment ranges from alchemy, prophecy and faith healing, to haunted houses, the Crusades, and the "influence of politics and religion on the hair and beard." The value of the book, like the account of Alexander the Oracle-Monger, lies in the remoteness of the frauds and delusions described. Many of the impostures do not have a contemporary ring and only weakly engage our passions: it becomes clear how people in other times were deceived. But after reading many such cases, we begin to wonder what the comparable contemporary versions are. People's feelings are as strong as they always were, and skepticism is probably as unfashionable today as in any other age. Accordingly, there ought to be bamboozles galore in contemporary society. And there are.

In Alexander's time, as in Mackay's, religion was the source of most accepted insights and prevailing world views. Those intent on duping the public often did so in religious language. This is, of course, still being done, as the testimony of penitent spiritualists and other late-breaking news amply attest. But in the past hundred years—whether for good or for ill—science has emerged in the popular mind as the primary means of penetrating the secrets of the universe, so we should expect many contemporary bamboozles to have a scientific ring. And they do.

Within the last century or so, many claims have been made at the edge or border of science—assertions that excite popular interest and, in many cases, that would be of profound scientific importance if only they were true. We will shortly examine a representative sampling of them. These claims are out of the ordinary, a break from the humdrum world, and often imply something hopeful: for example, that we have vast, untapped powers, or that unseen forces are about to save us from ourselves, or that there is a still unacknowledged pattern and harmony to the universe. Well, science does sometimes make such claims—as, for example, the realization that the hereditary information we pass from generation to generation is encoded in a single long molecule called DNA, in the discovery of universal gravitation or continental drift, in the tapping of nuclear energy, in research on the origin of life or on the early history of the universe. So if some additional claim is made—for example, that it is possible to float in the air unaided, by a special effort of will—what is so different about that? Nothing. Except for the matter of proof. Those who claim that levitation occurs have an obligation to demonstrate their contention before skeptics, under controlled conditions. The burden of proof is on them, not on those who might be dubious. Such claims are too important to think about carelessly. Many assertions about levitation have been made in the last hundred years, but motion pictures of well-illuminated people rising unassisted fifteen feet into the air have never been taken under conditions which exclude fraud. If levitation were possible, its scientific and, more generally, its human implications would be enormous. Those who make uncritical observations or fraudulent claims lead us into error and deflect us from the major human goal of understanding how the word works. It is for this reason playing fast and loose with the truth is a very serious matter.

ASTRAL PROJECTION

CONSIDER WHAT is sometimes called astral projection. Under conditions of religious ecstasy or hypnagogic sleep, or sometimes under the influence of a hallucinogen, people report the distinct sensation of stepping outside the body, leaving it, floating effortlessly to some other place in the room (often near the ceiling), and only at the end of the experience remerging with the body. If such a thing can actually happen, it is certainly of great importance; it implies something about the nature of human personality and even about the possibility of "life after death." Indeed, some people who have had near-death experiences, or who have been declared clinically dead and then revived, report similar sensations. But the fact that a sensation is reported does not mean that it occurred as claimed. There might, for example, be a common experience or wiring defect in human neuroanatomy that under certain circumstances always leads to the same illusion of astral projection. (See Chapter 25.)

There is a simple way to test astral projection. In your absence, have a friend place a book face up on a high and inaccessible shelf in the library. Then, if you ever have an astral projection experience, float to the book and read the title. When your body reawakens and you correctly announce what you have read, you will have provided some evidence for the physical reality of astral projection. But, of course, there must be no other way for you to know the title of the book, such as sneaking a peek when no one else is around, or being told by your friend or by someone your friend tells. To avoid the latter possibility, the experiment should be done "double blind"; that is, someone quite unknown to you who is entirely unaware of your existence must select and place the book and judge whether your answer is correct. To the best of my knowledge no demonstration of astral projection has ever been reported under such controlled circumstances with skeptics in attendance. I conclude that while astral projection is not excluded, there is little reason to believe in it. On the other hand, there is some evidence accumulated by Ian Stevenson, a University of Virginia psychiatrist, that young children in India and the Near East report in great detail a previous life in a moderately distant locale which they have never visited, while further inquiry demonstrates that a recently deceased person fits the child's description very well. But this is not an experiment performed under controlled conditions, and it is at least possible that the child has overheard or been given information about which the investigator is unaware. Stevenson's work is probably the most interesting of all contemporary research on "extrasensory perception."

SPIRIT RAPPING

IN UPSTATE NEW YORK in 1848 there lived two little girls, Margaret and Kate Fox, about whom marvelous stories were told. In their presence could be heard mysterious rapping noises, later understood to be coded messages from the spirit world: Ask the spirits anything—one rap signifies no, three raps signify yes. The Fox sisters became a sensation, embarked on nationwide tours organized by their elder sister, and became the focus of rapt attention from European intellectuals and literati such as Elizabeth Barrett Browning. The "manifestations" brought about by the Fox sisters are the origins of modern spiritualism, the belief that by some special effort of will a few gifted people are able to communicate with the spirits of the dead. Keene's associates owe a substantial debt to the Fox sisters.

Forty years after the first "manifestations," provoked by an uneasy conscience, Margaret Fox produced a signed confession. The raps were made—in a standing position with no apparent effort or movement—by cracking the toe and ankle joints, very much like cracking knuckles. "And that is the way we began. First, as a mere trick to frighten mother, and then, when so many people came to see us children, we were ourselves frightened, and for self-preservation forced to keep it up. No one suspected us of any trick because we were such young children. We were led on by my sister purposely and by mother unintentionally." The eldest sister, who organized their tours, seems to have been fully conscious of the fraud. Her motive was money.

The most instructive aspect of the Fox case is not that so many people were bamboozled; but rather that after the hoax was confessed, after Margaret Fox made a public demonstration, on the stage of a New York theater, of her "preternatural big toe," many who had been taken in still refused to acknowledge the fraud. They pretended that Margaret had been coerced into the confession by some rationalist Inquisition. People are rarely grateful for a demonstration of their credulity.

THE CARDIFF GIANT

IN 1869 THE FIGURE of a larger-than-life stone man was unearthed by a farmer "while digging a well" near the village of Cardiff in western New York. Clergymen and scientists alike asserted that it was a fossilized human being from ages past, perhaps a confirmation of the Biblical account: "There were giants in those days." Many commented on the detail of the figure, seemingly far finer than a mere artisan could have carved from stone. Why, there were even networks of tiny blue veins. But others were less impressed, including Andrew Dickson White, the first president of Cornell University, who declared it to be a pious fraud, and execrable sculpture to boot. A meticulous examination then revealed it to be of very recent origin, whereupon it emerged that the Cardiff Giant was merely a statue, a hoax engineered by George Hull of Binghamton, who described himself as "tobacconist, inventor, alchemist, atheist," a busy man. The "blue veins" were a natural pattern in the sculpted rock. The object of the deception was to fleece tourists.

But this uncomfortable revelation did not faze the American entrepreneur P. T. Barnum, who offered $60,000 for a three-month lease on the Cardiff Giant. When Barnum failed to secure it for traveling exhibition (the owners were making too much money to give it up), he simply had a copy made and exhibited *it,* to the awe of his customers and the enrichment of his pocketbook. The Cardiff Giant that most Americans have seen is this copy. Barnum exhibited a fake fake. The original is today languishing at the Farmer's Museum in Cooperstown, New York. Both Barnum and H. L. Mencken are said to have made the depressing observation that no one ever lost money by underestimating the intelligence of the American public. The remark has worldwide application. But the lack is not in intelligence, which is in plentiful supply; rather, the scarce commodity is systematic training in critical thinking.

CLEVER HANS, THE MATHEMATICAL HORSE

IN THE EARLY YEARS of the twentieth century there was a horse in Germany who could read, do mathematics and exhibit a deep knowledge of world political affairs. Or so it seemed. The horse was called Clever Hans. He was owned by Wilhelm von Osten, an elderly Berliner whose character was such, everyone said, that fraud was out of the question. Delegations of distinguished scientists viewed the equine marvel and pronounced it genuine. Hans would reply to mathematical problems put to him with coded taps of his foreleg, and would answer nonmathematical questions by nodding his head up and down or shaking it side to side in the conventional Western way. For example, someone would say, "Hans, how much is twice the square root of nine, less one?" After a moment's pause Hans would dutifully raise his right foreleg and tap five times. Was Moscow the capital of Russia? Head shake. How about St. Petersburg? Nod.

The Prussian Academy of Sciences sent a commission, headed by Oskar Pfungst, to take a closer look; Osten, who believed fervently in Hans's powers, welcomed the inquiry. Pfungst noticed a number of interesting regularities. Sometimes, the more difficult the question, the longer it took Hans to answer; or when Osten did not know the answer, Hans exhibited a comparable ignorance; or when Osten was out of the room, or when the horse was blindfolded, no correct answers were forthcoming. But other times Hans would get the right answer in a strange place, surrounded by skeptics, with Osten not only out of the room, but out of town.

The solution eventually became clear. When a mathematical question was put to Hans, Osten would become slightly tense, for fear Hans would make too few taps. When Hans, however, reached the correct number of taps, Osten unconsciously and imperceptibly nodded or relaxed—imperceptibly to virtually all human observers, but not to Hans, who was rewarded with a sugar cube for correct answers. Even teams of skeptics would watch Hans's foot as soon as the question was put and make gestural or postural responses when the horse reached the right answer. Hans was totally ignorant of mathematics, but very sensitive to unconscious nonverbal cues. Similar signs were unknowingly transmitted to the horse when verbal questions were posed. Clever Hans was aptly named; he was a horse who had conditioned one human being and discovered that other human beings he had never before met would provide him the needed cues. But despite the unambiguous nature of Pfungst's evidence, similar stories of counting, reading and politically sage horses, pigs and geese have continued to plague the gullible of many nations.*

PRECOGNITIVE DREAMS

ONE OF THE MOST striking apparent instances of extrasensory perception is the precognitive experience, when a person has a compelling perception of an imminent disaster, the death of a loved one, or a communication from a long-lost friend, and the predicted event then transpires. Many who have had such experiences report

* For example, Lady Wonder, a horse from Virginia, could answer questions by arranging lettered wood blocks with her nose. Since she also replied to queries posed privately to her owner, she was pronounced not only literate but telepathic by the parapsychologist J. B. Rhine (*Journal of Abnormal and Social Psychology, 23,* 449, 1929). The magician John Scarne found the owner would intentionally signal the horse with a whip as Lady Wonder moved her head over the blocks, preparatory to nudging them into words. The owner seemed to be out of the horse's field of view, but horses have excellent peripheral vision. Unlike Clever Hans, Lady Wonder was an accomplice in an intentional fraud.

that the emotional intensity of the precognition and its subsequent verification provide an overpowering sense of contact with another realm of reality. I have had such an experience myself. Many years ago I awoke in the dead of night in a cold sweat, with the certain knowledge that a close relative had suddenly died. I was so gripped with the haunting intensity of the experience that I was afraid to place a long-distance phone call, for fear that the relative would trip over the telephone cord (or something) and make the experience a self-fulfilling prophecy. In fact, the relative is alive and well, and whatever psychological roots the experience may have, it was not a reflection of an imminent event in the real world.

However, suppose the relative had in fact died that night. You would have had a difficult time convincing me that it was merely coincidence. But it is easy to calculate that if each American has such a premonitory experience a few times in his lifetime, the actuarial statistics alone will produce a few *apparent* precognitive events somewhere in America each year. We can calculate that this must occur fairly frequently, but to the rare person who dreams of disaster, followed rapidly by its realization, it is uncanny and awesome. Such a coincidence must happen to *someone* every few months. But those who experience a correct precognition understandably resist its explanation by coincidence.

After my experience I did not write a letter to an institute of parapsychology relating a compelling predictive dream which was not borne out by reality. That is not a memorable letter. But had the death I dreamt actually occurred, such a letter would have been marked down as evidence for precognition. The hits are recorded, the misses are not. Thus human nature unconsciously conspires to produce a biased reporting of the frequency of such events.

THESE CASES—Alexander the Oracle-Monger, Keene, astral projection, the Fox sisters, the Cardiff Giant, Clever Hans and precognitive dreams—are typical of claims made on the boundary or edge of science. An amazing assertion is made, something out of the ordinary, marvelous or awesome—or at least not tedious. It survives superficial scrutiny by lay people and, sometimes, more detailed study and more impressive endorsement by celebrities and scientists. Those who accept the validity of the assertion resist all attempts at conventional explanation. The most common correct explanations are of two sorts. One is conscious fraud, usually by those with a financial interest in the outcome, as with the Fox sisters and the Cardiff Giant. Those who accept the phenomena have been bamboozled. The other explanation often applies when the phenomena are uncommonly subtle and complex, when nature is more intricate than we have guessed, when deeper study is required for understanding; Clever Hans and many precognitive dreams fit this second explanation. Here, very often, we bamboozle ourselves.

I have chosen the foregoing cases for another reason. They are all closely involved with everyday life—human or animal behavior, evaluating the reliability of evidence, occasions for the exercise of common sense. None of these cases involve technological complexities or arcane theoretical developments. We do not need an advanced degree in physics, let us say, to have our skeptical hackles rise at the pretensions of modern spiritualists. Nevertheless, these hoaxes, impostures and misapprehensions have captivated millions. How much more dangerous and difficult to assess must be borderline claims at the edge of less familiar sciences—about cloning, say, or cosmic catastrophes or lost continents or unidentified flying objects?

I make a distinction between those who perpetrate and promote borderline belief systems and those who accept them. The latter are often taken by the novelty of the systems, and the feeling of insight and grandeur they provide. These are in fact scientific attitudes and scientific goals. It is easy to imagine extraterrestrial visitors who looked like human beings, and flew space vehicles and even airplanes like our own, and taught our ancestors civilization. This does not strain our imaginative powers overly and is sufficiently similar to familiar Western religious stories to seem comfortable. The search for Martian microbes of exotic biochemistry, or for interstellar radio messages from intelligent beings biologically very dissimilar is more difficult to grasp and not as comforting. The former view is widely purveyed and available; the latter much less so. Yet I think many of those excited by the idea of ancient astronauts are motivated by sincere scientific (and occasionally religious) feelings. There is a vast untapped popular interest in the deepest scientific questions. For many people, the shoddily thought out doctrines of borderline science are the closest approximation to comprehensible science readily available. The popularity of borderline science is a rebuke to the schools, the press and commercial television for their sparse, unimaginative and ineffective efforts at science education; and to us scientists, for doing so little to popularize our subject.

Advocates of ancient astronauts—the most notable being Erich von Däniken in his book *Chariots of the Gods?*—assert that there are numerous pieces of archaeological evidence that can be understood only by past contact by extraterrestrial civilizations with our ancestors. An iron pillar in India; a plaque in Palenque, Mexico; the pyramids of Egypt; the stone monoliths (all of which, according to Jacob Bronowski, resemble Benito Mussolini) on Easter Island; and the geometrical figures in Nazca, Peru, are all alleged to have been manufactured by or under the supervision of extraterrestrials. But in every case the artifacts in question have plausible and much simpler explanations. Our ancestors were no dummies. They may have lacked high technology, but they were as smart as we, and they sometimes combined dedication, intelligence and hard work to produce results that impress even us. The ancient-astronaut idea, interestingly, is popular among bureaucrats and politicians in the Soviet Union, perhaps because it preserves the old religious ideas in an acceptably modern scientific context. The most recent version of the ancient-astronaut story is the claim that the Dogon people in the Republic of Mali have an astronomical tradition concerning the star Sirius which they could only have acquired by contact with an alien civilization. This seems, in fact, to be the correct explanation, but it has nothing to do with astronauts, ancient or modern. (See Chapter 6.)

It is not surprising that pyramids have played a role in ancient-astronaut writings; ever since the Napoleonic invasions of Egypt impressed ancient Egyptian civilization on the consciousness of Europe, they have been the focus of a great deal of nonsense. Much has been written about supposed numerological information stored in the dimensions of the pyramids, especially the great pyramid of Gizeh, so that, for example, the ratio of height to width in certain units is said to be the time between Adam and Jesus in years. In one famous case a pyramidologist was observed filing a protuberance so that the observations and his speculations would be in better accord. The most recent manifestation of interest in pyramids is "pyramidology," the contention that we and our razor blades feel better and last longer inside pyramids than we and they do inside cubes. Maybe. I find living in cubical dwellings depressing, and for most of our history human beings did not live in such quarters. But the contentions of pyramidology, under appropriately controlled conditions, have never been verified. Again, the burden of proof has not been met.

The Bermuda Triangle "mystery" has to do with unexplained disappearances of ships and airplanes in a vast region of the ocean around Bermuda. The most reasonable explanation for these disappearances (when they actually occur; many of the alleged disappearances turn out simply never to have happened) is that the vessels sank. I once objected on a television program that it seemed strange for ships and airplanes to disappear mysteriously but never trains; to which the host, Dick Cavett, replied, "I can see you've never waited for the Long Island Railroad." As with the ancient-astronaut enthusiasts, the Bermuda Triangle advocates use sloppy scholarship and rhetorical questions. But they have not provided compelling evidence. They have not met the burden of proof.

Flying saucers, or UFOs, are well known to almost everyone. But seeing a strange light in the sky does not mean that we are being visited by beings from the planet Venus or a distant galaxy named Spectra. It might, for example, be an automobile headlight reflected off a high-altitude cloud, or a flight of luminescent insects, or an unconventional aircraft, or a conventional aircraft with unconventional lighting patterns, such as a high-intensity searchlight used for meteorological observations. There are also a number of cases—closer encounters with some highish index numeral—where one or two people claim to have been taken aboard an alien spaceship, prodded and probed with unconventional medical instruments, and released. But in these cases we have only the unsubstantiated testimony, no matter how heartfelt and seemingly sincere, of one or two people. To the best of my knowledge there are no instances out of the hundreds of thousands of UFO reports filed since 1947—not a single one—in which many people independently and reliably report a close encounter with what is clearly an alien spacecraft.

Not only is there an absence of good anecdotal evidence; there is no physical evidence either. Our laboratories are very sophisticated. A product of alien manufacture might readily be identified as such. Yet no one has ever turned up even a small fragment of an alien spacecraft that has passed any such physical test—much less the logbook of the starship captain. It is for these reasons that in 1977 NASA declined an invitation from the Executive Office of the President to undertake a serious investigation of UFO reports. When hoaxes and mere anecdotes are excluded, there seems to be nothing left to study.

Once I spied a bright, "hovering" UFO, and pointing it out to some friends in a restaurant, soon found myself in the midst of a throng of patrons, waitresses, cooks and proprietors milling about on the sidewalk, pointing up into the sky with fingers and forks, and making gasps of astonishment. People were somewhere between delighted and awestruck. But when I returned with a pair of binoculars which clearly showed the

UFO to be an unconventional aircraft (a NASA weather airplane, as it later turned out), there was uniform disappointment. Some felt embarrassed at the public exposure of their credulity. Others were simply disappointed at the evaporation of a good story, something out of the ordinary—a visitor from another world.

In many such cases we are not unbiased observers. We have an emotional stake in the outcome—perhaps merely because the borderline belief system, if true, makes the world a more interesting place; but perhaps because there is something there that strikes more deeply into the human psyche. If astral projection actually occurs, then it is possible for some thinking and perceiving part of me to leave my body and effortlessly travel to other places—an exhilarating prospect. If spiritualism is real, then my soul will survive the death of my body—possibly a comforting thought. If there is extrasensory perception, then many of us possess latent talents that need only be tapped to make us more powerful than we are. If astrology is right, then our personalities and destinies are intimately tied to the rest of the cosmos. If elves and goblins and fairies truly exist (there is a lovely Victorian picture book showing photographs of six-inch-high undraped ladies with gossamer wings conversing with Victorian gentlemen), then the world is a more intriguing place than most adults have been led to believe. If we are now being or in historical times have been visited by representatives from advanced and benign extraterrestrial civilizations, perhaps the human predicament is not so dire as it seems; perhaps the extraterrestrials will save us from ourselves. But the fact that these propositions charm or stir us does not guarantee their truth. Their truth depends only on whether the evidence is compelling; and my own, and sometimes reluctant, judgment is that compelling evidence for these and many similar propositions simply does not (at least as yet) exist.

What is more, many of these doctrines, if false, are pernicious. In simplistic popular astrology we judge people by one of twelve character types depending on their month of birth. But if the typing is false, we do an injustice to the people we are typing. We place them in previously collected pigeonholes and do not judge them for themselves, a typing familiar in sexism and racism.

The interest in UFOs and ancient astronauts seems at least partly the result of unfulfilled religious needs. The extraterrestrials are often described as wise, powerful, benign, human in appearance, and sometimes they are attired in long white robes. They are very much like gods and angels, coming from other planets rather than from heaven, using spaceships rather than wings. There is a little pseudoscientific overlay, but the theological antecedents are clear: in many cases the supposed ancient astronauts and UFO occupants are deities, feebly disguised and modernized, but easily recognizable. Indeed, a recent British survey suggests that more people believe in extraterrestrial visitations than in God.

Classical Greece was replete with stories in which the gods came down to Earth and conversed with human beings. The Middle Ages were equally rich in apparitions of saints and Virgins. Gods, saints and Virgins were all recorded repeatedly over centuries by people of the highest apparent reliability. What has happened? Where have all the Virgins gone? What has happened to the Olympian gods? Have these beings simply abandoned us in recent and more skeptical times? Or could these early reports reflect the superstition and credulity and unreliability of witnesses? And this suggests a possible social danger from the proliferation of UFO cultism: if we believe that benign extraterrestrials will solve our problems, we may be tempted to exert less than our full measure of effort to solve them ourselves—as has occurred in millennialist religious movements many times in human history.

All the really interesting UFO cases depend on believing that one or a few witnesses were not bamboozling or bamboozled. Yet the opportunity for deception in eyewitness accounts is breathtaking: (1) When a mock robbery is staged for a law school class, few of the students can agree on the number of intruders, their clothing, weapons or comments, the sequence of events or the time the robbery took. (2) Teachers are presented with two groups of children who have, unknown to them, tested equally well on all examinations. But the teachers are informed that one group is smart and the other dumb. The subsequent grades reflect that initial and erroneous assessment, independent of the performance of the students. Predispositions bias conclusions. (3) Witnesses are shown a motion picture of an automobile accident. They are then asked a series of questions such as "Did the blue car run the stop sign?" A week later, when questioned again, a large proportion of the witnesses claim to have seen a blue car—despite the fact that no remotely blue car is in the film. There seems to be a stage, shortly after an eyewitness event, in which we verbalize what we think we have seen and then forever after lock it into our memories. We are very vulnerable in that stage, and any prevailing beliefs—in Olympian gods or Christian saints or extraterrestrial astronauts, say—can unconsciously influence our eyewitness account.

Those skeptical of many borderline belief systems are not necessarily those afraid of novelty. For example, many of my colleagues and I are deeply interested in the possibility of life, intelligent or otherwise, on other planets. But we must be careful not to foist our wishes and fears upon the cosmos. Instead, in the usual scientific tradition, our objective is to find out what the answers really are, independent of our emotional predispositions. If we are alone, that is a truth worth knowing also. No one would be more delighted than I if intelligent extraterrestrials were visiting our planet. It would make my job enormously easier. Indeed, I have spent more time than I care to think about on the UFO and ancient astronaut questions. And public interest in these matters is, I believe, at least in part, a good thing. But our openness to the dazzling possibilities presented by modern science must be tempered by some hard-nosed skepticism. Many interesting possibilities simply turn out to be wrong. An openness to new possibilities and a willingness to ask hard questions are both required to advance our knowledge. And the asking of tough questions has an ancillary benefit: political and religious life in America, especially in the last decade and a half, has been marked by an excessive public credulity, an unwillingness to ask difficult questions, which has produced a demonstrable impairment in our national health. Consumer skepticism makes quality products. This may be why governments and churches and school systems do not exhibit unseemly zeal in encouraging critical thought. They know they themselves are vulnerable.

Professional scientists generally have to make a choice in their research goals. There are some objectives that would be very important if achieved, but that promise so small a likelihood of success that no one is willing to pursue them. (For many years this was the case in the search for extraterrestrial intelligence. The situation has changed mainly because advances in radio technology now permit us to construct enormous radio telescopes with sensitive receivers to pick up any messages that might be sent our way. Never before in human history was this possible.) There are other scientific objectives that are perfectly tractable but of entirely trivial significance. Most scientists choose a middle course. As a result, very few scientists actually plunge into the murky waters of testing or challenging borderline or pseudo-scientific beliefs. The chance of finding out something really interesting—except about human nature—seems small, and the amount of time required seems large. I believe that scientists should spend more time in discussing these issues, but the fact that a given contention lacks vigorous scientific opposition in no way implies that scientists think it is reasonable.

There are many cases where the belief system is so absurd that scientists dismiss it instantly but never commit their arguments to print. I believe this is a mistake. Science, especially today, depends upon public support. Because most people have, unfortunately, a very inadequate knowledge of science and technology, intelligent decision making on scientific issues is difficult. Some pseudoscience is a profitable enterprise, and there are proponents who not only are strongly identified with the issue in question but also make large amounts of money from it. They are willing to commit major resources to defending their contentions. Some scientists seem unwilling to engage in public confrontations on borderline science issues because of the effort required and the possibility that they will be perceived to lose a public debate. But it is an excellent opportunity to show how science works at its murkier borders, and also a way to convey something of its power as well as its pleasures.

There is stodgy immobility on both sides of the borders of the scientific enterprise. Scientific aloofness and opposition to novelty are as much a problem as public gullibility. A distinguished scientist once threatened to sic then Vice President Spiro T. Agnew on me if I persisted in organizing a meeting of the American Association for the Advancement of Science in which both proponents and opponents of the extraterrestrial-spacecraft hypothesis of UFO origins would be permitted to speak. Scientists offended by the conclusions of Immanuel Velikovsky's *Worlds in Collision* and irritated by Velikovsky's total ignorance of many well-established scientific facts successfully and shamefully pressured Velikovsky's publisher to abandon the book—which was then put out by another firm, much to its profit—and when I arranged for a second AAAS symposium to discuss Velikovsky's ideas, I was criticized by a different leading scientist who argued that any public attention, no matter how negative, could only aid Velikovsky's cause.

But these symposia were held, the audiences seemed to find them interesting, the proceedings were published, and now youngsters in Duluth or Fresno can find some books presenting the other side of the issue in their libraries. (See pages 71-72.) If science is presented poorly in schools and the media, perhaps some interest can be aroused by well-prepared, comprehensible public discussions at the edge of science. Astrology can be used for discussions of astronomy; alchemy for chemistry; Velikovskian catastrophism and lost continents such as Atlantis for geology; and spiritualism and Scientology for a wide range of issues in psychology and psychiatry.

There are still many people in the United States who believe that if a thing appears in print it must be true. Since so much undemonstrated speculation and rampant nonsense appears in books, a curiously distorted view of what is true emerges. I was amused to read—in the furor that followed the premature newspaper release of the contents of a book by H. R. Haldeman, a former presidential assistant and convicted felon—what the editor in chief of one of the largest publishing companies in the world had to say: "We believe a publisher has an obligation to check out the accuracy of certain controversial non-fiction works. Our procedure is to send the book out for an objective reading by an independent authority in the field." This is by an editor whose firm has in fact published some of the most egregious pseudoscience of recent decades. But books presenting the other side of the story are now becoming available, and in the section below I have listed a few of the more prominent pseudoscientific doctrines and recent attempts at their scientific refutation. One of the contentions criticized—that plants have emotional lives and musical preferences—had a brief flurry of interest a few years ago, including weeks of conversations with vegetables in Gary Trudeau's "Doonesbury" comic strip. As an epigraph to this chapter (on the death struggle of the snapdragon) shows, it is an old contention. Perhaps the only encouraging point is that it is being greeted more skeptically today than it was in 1926.

SOME RECENT BORDERLINE DOCTRINES
AND THEIR CRITIQUES

While many recent borderline doctrines are widely promoted, skeptical discussion and dissection of their fatal flaws are not so widely known. This table is a guide to some of these critiques.

Bermuda Triangle	*The Bermuda Triangle Mystery—Solved,* Laurence Kusche, Harper & Row, 1975
Spiritualism	*A Magician Among the Spirits,* Harry Houdini, Harper, 1924

	The Psychic Mafia, M. Lamar Keene, St. Martin's Press, 1976
Uri Geller	*The Magic of Uri Geller,* James Randi, Ballantine, 1975
Atlantis and other "lost continents"	*Legends of the Earth: Their Geologic Origins,* Dorothy B. Vitaliano, Indiana University Press, 1973
	Lost Continents, L. Sprague de Camp, Ballantine, 1975
UFOs	*UFOs Explained,* Philip Klass, Random House, 1974
	UFOs: A Scientific Debate, Carl Sagan and Thornton Page, eds., Norton, 1973
Ancient Astronauts	*The Space Gods Revealed: A Close Look at the Theories of Erich von Däniken,* Ronald Story, Harper & Row, 1976
	The Ancient Engineers, L. Sprague de Camp, Ballantine, 1973
Velikovsky: *Worlds in Collision*	*Scientists Confront Velikovsky,* Donald Goldsmith, ed., Cornell University Press, 1977
The Emotional Lives of Plants	"Plant 'Primary Perception,' " K. A. Horowitz and others, *Science,* 189: 478–480 (1975)

A FEW YEARS AGO a committee of scientists, magicians and others was organized to provide some focus for skepticism on the border of science. This nonprofit organization is called "The Committee for the Scientific Investigation of Claims of the Paranormal" and is at 923 Kensington Avenue, Buffalo, N.Y. 14215. It is beginning to do some useful work, including in its publications the latest news on the confrontation between the rational and the irrational—a debate that goes back to the encounters between Alexander the Oracle-Monger and the Epicureans, who were the rationalists of his day. The committee has also made official protests to the networks and the Federal Communications Commission about television programs on pseudoscience that are particularly uncritical. An interesting debate has gone on within the committee between those who think that all doctrines that smell of pseudoscience should be combated and those who believe that each issue should be judged on its own merits, but that the burden of proof should fall squarely on those who make the proposals. I find myself very much in the latter camp. I believe that the extraordinary should certainly be pursued. But extraordinary claims require extraordinary evidence.

Scientists are, of course, human. When their passions are excited they may abandon temporarily the ideals of their discipline. But these ideals, the scientific method, have proved enormously effective. Finding out the way the world really works requires a mix of hunches, intuition and brilliant creativity; it also requires skeptical scrutiny of every step. It is the tension between creativity and skepticism that has produced the stunning and unexpected findings of science. In my opinion the claims of borderline science pall in comparison with hundreds of recent activities and discoveries in real science, including the existence of two semi-independent brains within each human skull; the reality of black holes; continental drift and collisions; chimpanzee language; massive climatic changes on Mars and Venus; the antiquity of the human species; the search for extraterrestrial life; the elegant self-copying molecular architecture that controls our heredity and evolution; and observational evidence on the origin, nature and fate of the universe as a whole.

But the success of science, both its intellectual excitement and its practical application, depend upon the self-correcting character of science. There must be a way of testing any valid idea. It must be possible to reproduce any valid experiment. The character or beliefs of the scientist are irrelevant; all that matters is whether the evidence supports his contention. Arguments from authority simply do not count; too many authorities have been mistaken too often. I would like to see these very effective scientific modes of thought communicated by the schools and the media; and it would certainly be an astonishment and delight to see them introduced into politics. Scientists have been known to change their minds completely and publicly when presented with new evidence or new arguments. I cannot recall the last time a politician displayed a similar openness and willingness to change.

Many of the belief systems at the edge or fringe of science are not subject to crisp experimentation. They are anecdotal, depending entirely on the validity of eyewitnesses who, in general, are notoriously unreliable. On the basis of past performance most such fringe systems will turn out to be invalid. But we cannot reject out of hand, any more than we can accept at face value, all such contentions. For example, the idea that large rocks can drop from the skies was considered absurd by eighteenth-century scientists; Thomas Jefferson remarked about one such account that he would rather believe that two Yankee scientists lied than that stones fell from the heavens. Nevertheless, stones do fall from the heavens. They are called meteorites, and our preconceptions have no bearing on the truth of the matter. But the truth was established only by a careful analysis of dozens of independent witnesses to a common meteorite fall, supported by a great body of physical evidence, including meteorites recovered from the eaves of houses and the furrows of plowed fields.

Prejudice means literally pre-judgment, the rejection of a contention out of hand, before examining the evidence. Prejudice is the result of powerful emotions, not of sound reasoning. If we wish to find out the truth of a matter we must approach the question with as nearly open a mind as we can, and with a deep awareness of our own limitations and predispositions. On the other hand, if after carefully and openly examining the evidence, we reject the proposition, that is not prejudice. It might be called "post-judice." It is certainly a prerequisite for knowledge.

Critical and skeptical examination is the method used in everyday practical matters as well as in science. When buying a new or used car, we think it prudent to insist on written warranties, test drives and checks of particular parts. We are very careful about car dealers who are evasive on these points. Yet the practitioners of many borderline beliefs are offended when subjected to similarly close scrutiny. Many who claim to have extrasensory perception also claim that their abilities decline when they are carefully watched. The magician Uri Geller is happy to warp keys and cutlery in the vicinity of scientists—who, in their confrontations with nature, are used to an adversary who fights fair; but is greatly affronted at the idea of performances before an audience of skeptical magicians—who, understanding human limitations, are themselves able to perform similar effects by sleight of hand. Where skeptical observation and discussion are suppressed, the truth is hidden. The proponents of such borderline beliefs, when criticized, often point to geniuses of the past who were ridiculed. But the fact that some geniuses were laughed at does not imply that all who are laughed at are geniuses. They laughed at Columbus, they laughed at Fulton, they laughed at the Wright brothers. But they also laughed at Bozo the Clown.

The best antidote for pseudoscience, I firmly believe, is science:

■ There is an African fresh-water fish that is blind. It generates a standing electric field, through perturbations in which it distinguishes between predators and prey and communicates in a fairly elaborate electrical language with potential mates and other fish of the same species. This involves an entire organ system and sensory capability completely unknown to pretechnological human beings.

■ There is a kind of arithmetic, perfectly reasonable and self-contained, in which two times one does not equal one times two.

■ Pigeons—one of the least prepossessing animals on Earth—are now found to have a remarkable sensitivity to magnetic-field strengths as small as one hundred thousandth that of the Earth's magnetic dipole. Pigeons evidently use this sensory capability for navigation and sense their surroundings by their magnetic signatures: metal gutters, electrical power lines, fire escapes and the like—a sensory modality glimpsed by no human being who ever lived.

- Quasars seem to be explosions of almost unimaginable violence in the hearts of galaxies which destroy millions of worlds, many of them perhaps inhabited.
- In an East African volcanic ash flow 3.5 million years old there are footprints—of a being about four feet high with a purposeful stride that may be the common ancestor of apes and men. Nearby are the prints of a knuckle-walking primate corresponding to no animal yet discovered.
- Each of our cells contains dozens of tiny factories called mitochondria which combine our food with molecular oxygen in order to extract energy in convenient form. Recent evidence suggests that billions of years ago the mitochondria were free organisms which have slowly evolved into a mutually dependent relation with the cell. When many-celled organisms arose, the arrangement was retained. In a very real sense, then, we are not a single organism, but an array of about ten trillion beings and not all of the same kind.
- Mars has a volcano almost 80,000 feet high which was constructed about a billion years ago. An even larger volcano may exist on Venus.
- Radio telescopes have detected the cosmic blackbody background radiation, the distant echo of the event called the Big Bang. The fires of creation are being observed today.

I could continue such a list almost indefinitely. I believe that even a smattering of such findings in modern science and mathematics is far more compelling and exciting than most of the doctrines of pseudoscience, whose practitioners were condemned as early as the fifth century B.C. by the Ionian philosopher Heraclitus as "night-walkers, magicians, priests of Bacchus, priestesses of the wine-vat, mystery-mongers." But science is more intricate and subtle, reveals a much richer universe, and powerfully evokes our sense of wonder. And it has the additional and important virtue—to whatever extent the word has any meaning—of being true.

CHAPTER 6

WHITE DWARFS AND LITTLE GREEN MEN

◆◆◆

No testimony is sufficient to establish a miracle,
unless . . . its falsehood would be more
miraculous than the fact
which it endeavors to establish.

DAVID HUME,
Of Miracles

HUMANITY HAS already achieved interstellar spaceflight. With a gravitational assist from the planet Jupiter, the Pioneer 10 and 11 and the Voyager 1 and 2 spacecraft have been boosted into trajectories that will leave the solar system for the realm of the stars. They are very slow-moving spacecraft despite the fact that they are the fastest objects ever launched by our species. They will take tens of thousands of years to travel typical interstellar distances. Unless some special effort is made to redirect them, they will never enter another planetary system in all the tens of billions of years of future history of the Milky Way Galaxy. The star-to-star distances are too large. They are doomed to wander forever in the dark between the stars. But even so, these spacecraft have messages attached to them for the remote contingency that at some future time, alien beings might intercept the spacecraft and wonder about the beings who launched them on these prodigious journeys.*

If we are capable of such constructions at our comparatively backward technological state, might not a civilization thousands or millions of years more advanced than ours, on a planet of another star, be capable of fast and directed interstellar travel? Interstellar spaceflight is time-consuming, difficult and expensive for us, and perhaps also for other civilizations with substantially greater resources than ours. But it surely would be unwise to contend that conceptually novel approaches to the physics or engineering of interstellar spaceflight will not be discovered by us sometime in the future. It is evident that for economy, efficiency and convenience, interstellar radio transmission is much superior to interstellar spaceflight, and this is the reason that our own efforts have concentrated strongly on radio communication. But radio communication is clearly inappropriate for contact with a pretechnological society or species. No matter how clever or powerful the transmission, no such radio message would have been received or understood on Earth before the present century. And there has been life on our planet for about four billion years, human beings for several million, and civilization for perhaps ten thousand.

It is not inconceivable that there is a kind of Galactic Survey, established by cooperating civilizations on many planets throughout the Milky Way Galaxy, which keeps an eye (or some equivalent organ) on emerging planets and seeks out undiscovered worlds. But the solar system is very far from the center of the Galaxy and could well have eluded such searches. Or survey ships may

---

* A detailed discussion of the Pioneer 10 and 11 plaque can be found in my book *The Cosmic Connection* (New York, Doubleday, 1973); and the phonograph records aboard Voyager 1 and 2 are comprehensively described in *Murmurs of Earth: The Voyager Interstellar Record* (New York, Random House, 1978).

come here, but only every ten million years, say—with none having arrived during historical times. However, it is also possible that a few survey teams have arrived recently enough in human history for their presence to have been noted by our ancestors, or even for human history to have been affected by the contact.

The Soviet astrophysicist I. S. Shklovskii and I discussed this possibility in our book, *Intelligent Life in the Universe,* in 1966. We examined a range of artifacts, legends and folklore from many cultures and concluded that not one of these cases provided even moderately convincing evidence for extraterrestrial contact. There are always more plausible alternative explanations based on known human abilities and behavior. Among the cases discussed were a number later accepted by Erich von Däniken and other uncritical writers as valid evidence for extraterrestrial contact: Sumerian legends and astronomical cylinder seals; the Biblical stories of the Slavonic Enoch and of Sodom and Gomorrah; the Tassili frescoes in North Africa; the machined metal cube allegedly found in ancient geological sediments and said to be displayed in a museum in Austria; and so on. Over the years I have continued to look as deeply as I am able into such stories and have found very few that require more than passing attention.

In the long litany of "ancient astronaut" pop archaeology, the cases of apparent interest have perfectly reasonable alternative explanations, or have been misreported, or are simple prevarications, hoaxes and distortions. This description applies to arguments about the Piri Reis map, the Easter Island monoliths, the heroic drawings on the plains of Nazca, and various artifacts from Mexico, Uzbekistan and China.

And yet, it would be so easy for an advanced extraterrestrial civilization to leave a completely unambiguous calling card of their visit. For example, many nuclear physicists believe that there is an "island of stability" of atomic nuclei, near a hypothetical superheavy atom with about 114 protons and about 184 neutrons. All chemical elements heavier than uranium (with 238 protons and neutrons in its nucleus) spontaneously decay in cosmically short periods of time. But there is reason to think that the binding between protons and neutrons is such that stable elements would be produced if nuclei having about 114 protons and 184 neutrons could be constructed. Such a construction is just beyond our present technology, and clearly beyond the technology of our ancestors. A metal artifact containing such elements would be unambiguous evidence of an advanced extraterrestrial civilization in our past. Or consider the element technetium, whose most stable form has 99 protons and neutrons. Half of it radioactively decays to other elements in about 200,000 years, half of the remainder is gone in another 200,000 years, and so on. As a result, any technetium formed by stars with the other elements billions of years ago must all be gone by now. Thus, terrestrial technetium can only be of artificial origin, as its very name indicates. A technetium artifact could have only one meaning. Similarly, there are common elements on Earth that are immiscible; for example, aluminum and lead. If you melt them together, the lead, being considerably heavier, sinks to the bottom. The aluminum floats to the top. However, in the zero g conditions of spaceflight there is no gravity in the melt to pull the heavier lead down, and exotic alloys such as Al/Pb can be produced. One of the objectives of NASA's early Shuttle missions will be to test out such alloying techniques. Any message written on an aluminum/lead alloy and retrieved from an ancient civilization would certainly commend itself to our attention today.

It is also possible that the content rather than the material of the message would clearly point to a science or technology beyond the abilities of our ancestors: for example, a vector calculus rendition of Maxwell's equations (with or without magnetic monopoles), or a graphical representation of the Planck black-body distribution for several different temperatures, or a derivation of the Lorentz transformation of special relativity. Even if the ancient civilization could not understand such writings, they might revere them as holy. But no cases of this sort have emerged—despite what is clearly a profitable market for tales of ancient or modern extraterrestrial astronauts. There have been debates on the purity of magnesium samples from purported crashed UFOs, but their purity was within the competence of American technology at the time of the incident. A supposed star map said to be retrieved (from memory) from the interior of a flying saucer does not, as alleged, resemble the relative positions of the nearest stars like the Sun; in fact, a close examination shows it to be not much better than the "star map" which would be produced if you took an old-fashioned quill pen and splattered a few blank pages with ink spots. With one apparent exception, there are no stories sufficiently detailed to dispose of other explanations and sufficiently accurate to portray correctly modern physics or astronomy to a prescientific or pretechnical people. The one exception is the remarkable mythology surrounding the star Sirius that is held by the Dogon people of the Republic of Mali.

There are at most a few hundred thousand Dogon alive today, and they have been studied intensively by anthropologists only since the 1930s. There are some elements of their mythology that are reminiscent of the legends of the ancient Egyptian civilization, and some anthropologists have assumed a weak Dogon cultural connection with ancient Egypt. The helical risings of Sirius were central to the Egyptian calendar and used to predict the inundations of the Nile. The most striking aspects of Dogon astronomy have been recounted by Marcel Griaule, a French anthropologist working in the 1930s and 1940s. While there is no reason to doubt Griaule's account, it is important to note that there is no earlier Western record of these remarkable Dogon folk beliefs and that all the information has been funneled through Griaule. The story has recently been popularized by a British writer, R. K. G. Temple.

In contrast to almost all prescientific societies, the Dogon hold that the planets as well as the Earth rotate about their axes and revolve about the Sun. This is a conclusion that can, of course, be achieved without high technology, as Copernicus demonstrated, but it is a very rare insight among the peoples of the Earth. It was taught, however, in ancient Greece by Pythagoras and by Philolaus, who perhaps held, in Laplace's words, "that the planets were inhabited and that the stars were suns, disseminated in space, being themselves centers of planetary systems." Such teachings, among a wide variety of contradictory ideas, might be just an inspired lucky guess.

The ancient Greeks believed there were only four elements—earth, fire, water and air—from which all else was constructed. Among the pre-Socratic philosophers there were those who made special advocacy for each one of these elements. If it had later turned out that the universe was indeed made more of one of these elements than another, we should not attribute remarkable prescience to the pre-Socratic philosopher who made the proposal. One of them was bound to be right on statistical grounds alone. In the same way, if we have several hundred or several thousand cultures, each with its own cosmology, we should not be astounded if, every now and then, purely by chance, one of them proposes an idea that is not only correct but also impossible for them to have deduced.

But, according to Temple, the Dogon go further. They hold that Jupiter has four satellites and that Saturn is encircled by a ring. It is perhaps possible that individuals of extraordinary eyesight under superb seeing conditions could, in the absence of a telescope, have observed the Galilean satellites of Jupiter and the rings of Saturn. But this is at the bare edge of plausibility. Unlike every astronomer before Kepler, the Dogon are said to depict the planets moving correctly in elliptical, not circular, orbits.

More striking still is the Dogon belief about Sirius, the brightest star in the sky. They contend that it has a dark and invisible companion star which orbits Sirius (and, Temple says, in an elliptical orbit) once every fifty years. They state that the companion star is very small and very heavy, made of a special metal called "Sagala" which is not found on Earth.

The remarkable fact is that the visible star, Sirius A,

does have an extraordinary dark companion, Sirius B, which orbits it in an elliptical orbit once each 50.04 $\pm$0.09 years. Sirius B is the first example of a white dwarf star discovered by modern astrophysics. Its matter is in a state called "relativistically degenerate," which does not exist on Earth, and since the electrons are not bound to the nuclei in such degenerate matter, it can properly be described as metallic. Since Sirius A is called the Dog Star, Sirius B has sometimes been dubbed "The Pup."

At first glance the Sirius legend of the Dogon seems to be the best candidate evidence available today for past contact with an advanced extraterrestrial civilization. As we begin a closer look at this story, however, let us remember that the Dogon astronomical tradition is purely oral, that it dates with certainty only from the 1930s and that the diagrams are written with sticks in sand. (Incidentally, there is some evidence that the Dogon like to frame pictures with an ellipse, and that Temple may be mistaken about the claim that the planets and Sirius B move in elliptical orbits in Dogon mythology.)

When we examine the full body of Dogon mythology we find a very rich and detailed structure of legend—much richer, as many anthropologists have remarked, than those of their near geographical neighbors. Where there is a rich array of legends there is, of course, a greater chance of an accidental correspondence of one of the myths with a finding of modern science. A very spare mythology is much less likely to make such an accidental concordance. But when we examine the rest of Dogon mythology, do we find other cases hauntingly reminiscent of some unexpected findings in modern science?

The Dogon cosmogony describes how the Creator examined a plaited basket, round at the mouth and square at the bottom. Such baskets are still in use in Mali today. The Creator up-ended the basket and used it as a model for the creation of the world—the square base represents the sky and the round mouth the Sun. I must say that this account does not strike me as a remarkable anticipation of modern cosmological thinking. In the Dogon representation of the creation of the Earth, the Creator implants in an egg two pairs of twins, each pair comprised of a male and a female. The twins are intended to mature within the egg and fuse to become a single and "perfect" androgynous being. The Earth originates when one of the twins breaks from the egg before maturation, whereupon the Creator sacrifices the other twin in order to maintain a certain cosmic harmony. This is a variegated and interesting mythology, but it does not seem to be qualitatively different from many of the other mythologies and religions of humanity.

The hypothesis of a companion star to Sirius might have followed naturally from the Dogon mythology, in which twins play a central role, but there does not seem to be any explanation this simple about the period and density of the companion of Sirius. The Dogon Sirius myth is too close to modern astronomical thinking and too precise quantitatively to be attributed to chance. Yet there it sits, immersed in a body of more or less standard prescientific legend. What can the explanation be? Is there any chance that the Dogon or their cultural ancestors might actually have been able to see Sirius B and observe its period around Sirius A?

White dwarfs such as Sirius B evolve from stars called red giants, which are very luminous and, it will be no surprise to hear, red. Ancient writers of the first few centuries A.D. actually described Sirius as red—certainly not its color today. In a conversation piece by Horace called "Hoc Quoque Tiresia" (How to Get Rich Quickly) there is a quotation from an unspecified earlier work that says: "The red dog star's heat split the speechless statues." As a result of these less than compelling ancient sources there has been a slight temptation among astrophysicists to consider the possibility that the white dwarf Sirius B was a red giant in historical times and visible with the naked eye, completely swamping the light of Sirius A. In that case perhaps there was a slightly later time in the evolution of Sirius B when its brightness was comparable to that of

Sirius A, and the relative motion of the two stars about each other could be discerned with the unaided eye. But the best recent information from the theory of stellar evolution suggests that there simply is not enough time for Sirius B to have reached its present white dwarf state if it had been a red giant a few centuries before Horace. What is more, it would seem extraordinary that no one except the Dogon noticed these two stars circling each other every fifty years, each alone being one of the brightest stars in the sky. There was an extremely competent school of observational astronomers in Mesopotamia and in Alexandria in the preceding centuries—to say nothing of the Chinese and Korean astronomical schools—and it would be astonishing if they had noticed nothing.* Is our only alternative, then, to believe that representatives of an extraterrestrial civilization have visited the Dogon or their ancestors?

The Dogon have knowledge impossible to acquire without the telescope. The straightforward conclusion is that they had contact with an advanced technical civilization. The only question is, which civilization—extraterrestrial or European? Far more credible than an ancient extraterrestrial educational foray among the Dogon might be a comparatively recent contact with scientifically literate Europeans who conveyed to the Dogon the remarkable European myth of Sirius and its white dwarf companion, a myth that has all the superficial earmarks of a splendidly inventive tall story. Perhaps the Western contact came from a European visitor to Africa, or from the local French schools, or perhaps from contacts in Europe by West Africans inducted to fight for the French in World War I.

The likelihood that these stories arise from contact with Europeans rather than extraterrestrials has been increased by a recent astronomical finding: a Cornell

* The ancient Egyptian phrase for the planet Mars translates to "the red Horus," Horus being the imperial falcon deity. Thus Egyptian astronomy noted remarkable coloration in celestial objects. But the description of Sirius mentions nothing notable about its color.

University research team led by James Elliot, employing a high-altitude airborne observatory over the Indian Ocean, discovered in 1977 that the planet Uranus is surrounded by rings—a finding never hinted at by ground-based observations. Advanced extraterrestrial beings viewing our solar system upon approach to Earth would have little difficulty discovering the rings of Uranus. But European astronomers in the nineteenth and early twentieth centuries would have had nothing to say in this regard. The fact that the Dogon do not talk of another planet beyond Saturn with rings suggests to me that their informants were European, not extraterrestrial.

In 1844 the German astronomer F. W. Bessel discovered that the long-term motion of Sirius itself (Sirius A) was not straight but, rather, wavy against the background of more distant stars. Bessel proposed that there was a dark companion to Sirius whose gravitational influence was producing the observed sinusoidal motion. Since the period of the wiggle was fifty years, Bessel deduced that the dark companion had a fifty-year period in the joint motion of Sirius A and B about their common center of mass.

Eighteen years later Alvan G. Clark, during the testing of a new 18½-inch refracting telescope, accidentally discovered the companion, Sirius B, by direct visual observation. From the relative motions, Newtonian gravitational theory permits us to estimate the masses of Sirius A and B. The companion turns out to have a mass just about the same as the Sun's. But Sirius B is almost ten thousand times fainter than Sirius A, even though their masses are about the same and they are just the same distance from the Earth. These facts can be reconciled only if Sirius B has a much smaller radius or a much lower temperature. But in the late nineteenth century it was believed by astronomers that stars of the same mass had approximately the same temperature, and by the turn of the century it was widely held that the temperature of Sirius B was not remarkably low. Spectroscopic observations by Walter S. Adams in 1915 confirmed this contention. Hence,

Sirius B must be very small. We know today that it is only as big as the Earth. Because of its size and color it is called a white dwarf. But if Sirius B is much smaller than Sirius A, its density must be very much greater. Accordingly, the concept of Sirius B as an extremely dense star was widely held in the first few decades of this century.

The peculiar nature of the companion of Sirius was extensively reported in books and in the press. For example, in Sir Arthur Stanley Eddington's book *The Nature of the Physical World,* we read: "Astronomical evidence seems to leave practically no doubt that in the so-called *white dwarf* stars the density of matter far transcends anything of which we have terrestrial experience; in the Companion of Sirius, for example, the density is about a ton to the cubic inch. This condition is explained by the fact that the high temperature and correspondingly intense agitation of the material breaks up (ionises) the outer electron system of the atoms, so that the fragments can be packed much more closely together." Within a year of its 1928 publication, this book saw ten reprintings in English. It was translated into many languages, including French. The idea that white dwarfs were made of electron degenerate matter had been proposed by R. H. Fowler in 1925 and quickly accepted. On the other hand, the proposal that white dwarfs were made of "relativistically degenerate" matter was first made in the period 1934 to 1937, in Great Britain, by the Indian astrophysicist S. Chandrasekhar; the idea was greeted with substantial skepticism by astronomers who had not grown up with quantum mechanics. One of the most vigorous skeptics was Eddington. The debate was covered in the scientific press and was accessible to the intelligent layman. All this was occurring just before Griaule encountered the Dogon Sirius legend.

In my mind's eye I picture a Gallic visitor to the Dogon people, in what was then French West Africa, in the early part of this century. He may have been a diplomat, an explorer, an adventurer or an early anthropologist. Such people—for example, Richard Francis Burton—were in West Africa many decades earlier. The conversation turns to astronomical lore. Sirius is the brightest star in the sky. The Dogon regale the visitor with their Sirius mythology. Then, smiling politely, expectantly, they inquire of their visitor what *his* Sirius myths might be. Perhaps he refers before answering to a well-worn book in his baggage. The white dwarf companion of Sirius being a current astronomical sensation, the traveler exchanges a spectacular myth for a routine one. After he leaves, his account is remembered, retold, and eventually incorporated into the corpus of Dogon mythology—or at least into a collateral branch (perhaps filed under "Sirius myths, bleached peoples' account"). When Marcel Griaule makes mythological inquiries in the 1930s and 1940s, he has his own European Sirius myth played back to him.

THIS FULL-CYCLE RETURN of a myth to its culture of origin through an unwary anthropologist might sound unlikely if there were not so many examples of it in anthropological lore. I here recount a few cases:

In the first decade of the twentieth century a neophyte anthropologist was collecting accounts of ancient traditions from Native American populations in the Southwest. His concern was to write down the traditions, almost exclusively oral, before they vanished altogether. The young Native Americans had already lost appreciable contact with their heritage, and the anthropologist concentrated on elderly members of the tribe. One day he found himself sitting outside a hogan with an aged but lively and cooperative informant.

"Tell me about the ceremonies of your ancestors at the birth of a child."

"Just one moment."

The old Indian slowly shuffled into the darkened depths of the hogan. After a fifteen-minute interval he reappeared with a remarkably useful and detailed description of postpartum ceremonials, including rituals connected with breach presentation, afterbirth, umbilical cord, first breath and first cry. Encouraged and writing feverishly, the anthropologist systematically went through the full list of rites of passage, including puberty, marriage, childbearing and death. In each case the informant disappeared into the hogan only to emerge a quarter of an hour later with a rich set of answers. The anthropologist was astonished. Could, he wondered, there be a yet older informant, perhaps infirm and bedridden, within the hogan? Eventually he could resist no longer and summoned the courage to ask his informant what he did at each retreat into the hogan. The old man smiled, withdrew for the last time, and returned clutching a well-thumbed volume of the *Dictionary of American Ethnography*, which had been compiled by anthropologists in the previous decade. The poor white man, he must have thought, is eager, well-meaning and ignorant. He does not have a copy of this marvelous book which contains the traditions of my people. I shall tell him what it says.

My other two stories recount the adventures of an extraordinary physician, Dr. D. Carleton Gajdusek, who for many years has studied kuru, a rare viral disease, among the inhabitants of New Guinea. For this work he was the recipient of the 1976 Nobel Prize for Medicine. I am grateful to Dr. Gajdusek for taking the trouble to check my memory of his stories, which I first heard from him many years ago. New Guinea is an island on which mountainous terrain separates—in a manner similar to but more completely than the mountains of ancient Greece—one valley people from another. As a result there is a great profusion and variety of cultural traditions.

In the spring of 1957 Gajdusek and Dr. Vincent Zigas, a medical officer with the Public Health Service of what was then called the Territory of Papua and New Guinea, traveled with an Australian administrative patrol officer from the Purosa Valley through the ranges of the South Fore cultural and linguistic-group region to the village of Agakamatasa on an exploratory visit into "uncontrolled territory." Stone implements were still in use, and there remained a tradition of cannibalism within one's own living group. Gajdusek and his party found cases of kuru, which is spread by cannibalism (but most often not through the digestive tract), in this most remote of the South Fore villages. They decided to spend a few days, moving into one of the large and traditional *wa'e*, or men's houses (the music from one of which, incidentally, was sent to the stars on the Voyager phonograph record). The windowless, low-doored, smoky thatched house was partitioned so that the visitors could neither stand erect nor stretch out. It was divided into many sleeping compartments, each with its own small fire, around which men and boys would huddle in groups to sleep and keep warm during the cold nights at an elevation of more than 6,000 feet, an altitude higher than Denver. To accommodate their visitors, the men and boys gleefully tore out the interior structure of half of the ceremonial men's house, and during two days and nights of pouring rain Gajdusek and his companions were housebound on a high, windswept, cloud-covered ridge. The young Fore initiates wore bark strands braided into their hair, which was covered with pig grease. They wore huge nose pieces, the penises of pigs as armbands, and the genitalia of opossums and tree-climbing kangaroos as pendants around their necks.

The hosts sang their traditional songs all through the first night and on through the following rainy day. In return, "to enhance our rapport with them," as Gajdusek says, "we began to sing songs in exchange—among them such Russian songs as 'Otchi chornye,' and 'Moi kostyor v tumane svetit' . . ." This was received very well, and the Agakamatasa villagers requested many dozens of repetitions in the smoky South Fore longhouse to the accompaniment of the driving rainstorm.

Some years later Gajdusek was engaged in the collection of indigenous music in another part of the South Fore region and asked a group of young men to run through their repertoire of traditional songs. To Gajdusek's amazement and amusement, they produced a somewhat altered but still clearly recognizable version of "Otchi chornye." Many of the singers apparently thought the song traditional, and later still Gajdusek found the song imported even farther afield, with none of the singers having any idea of its source.

We can easily imagine some sort of world ethnomusicology survey coming to an exceptionally obscure part of New Guinea and discovering that the natives had a traditional song which sounded in rhythm, music and words remarkably like "Otchi chornye." If they were to believe that no previous contact of Westerners with these people had occurred, a great mystery could be posited.

Later that same year Gajdusek was visited by several Australian physicians, eager to understand the remarkable findings about the transmission of kuru from patient to patient by cannibalism. Gajdusek described the theories of the origin of many diseases held by the Fore people, who did *not* believe that illnesses were caused by the spirits of the dead or that malicious deceased relatives, jealous of the living, inflicted disease on those of their surviving kinsmen who offended them, as the pioneering anthropologist Bronislaw Malinowski had recounted for the coastal peoples of Melanesia. Instead, the Fore attributed most diseases to malicious sorcery, which any offended and avenging male, young or old, could execute without the aid of specially trained sorcerers. There was a special sorcery explanation for kuru, but also for chronic lung disease, leprosy, yaws, and so on. These beliefs had been long-established and firmly held, but as the Fore people witnessed yaws yielding entirely to the penicillin injections of Gajdusek and his group, they quickly agreed that the sorcery explanation of yaws was in error and abandoned it; it has never resurfaced in subsequent years. (I wish Westerners would be as quick to abandon obsolete or erroneous social ideas as the Fore of New Guinea.) Modern treatment of leprosy caused its sorcery explanation to disappear as well, although more slowly, and the Fore people today laugh at these backward early opinions on yaws and leprosy. But the traditional views on the origin of kuru have maintained themselves, since the Westerners have been unable to cure or explain, in a manner satisfactory to them, the origin and nature of this disease. Thus, the Fore people remain intensely skeptical of Western explanations for kuru and retain firmly their view that malicious sorcery is the cause.

One of the Australian physicians, visiting an adjacent village with one of Gajdusek's native informants as translator, spent the day examining kuru patients and independently acquiring information. He returned the same evening to inform Gajdusek that he was mistaken about people not believing in the spirits of the dead as the cause of disease, and that he was further in error in holding that they had abandoned the idea of sorcery as the cause of yaws. The people held, he continued, that a dead body could become invisible and that the unseen spirit of the dead person could enter the skin of a patient at night through an imperceptible break, and induce yaws. The Australian's informant had even sketched with a stick in the sand the appearance of these ghostly beings. They carefully drew a circle and a few squiggly lines within. Outside the circle, they explained, it was black; inside the circle, bright—a sand portrait of malevolent and pathogenic spirits.

Upon inquiry of the young translator, Gajdusek discovered that the Australian physician had conversed with some of the older men of the village who were well known to Gajdusek and who were often his house and laboratory guests. They had attempted to explain that the shape of the "germ" producing yaws was spiral—the spirochete form they had seen many times through Gajdusek's dark-field microscope. They had to admit it was invisible—it could be seen only through the microscope—and when pressed by the Australian physician on whether this "represented" the dead person, they had to admit that Gajdusek had stressed that it could be caught from close contact with yaws lesions, as, for example, by sleeping with a person with yaws.

I can well remember the first time I looked through a microscope. After focusing my eyes up near the ocular only to examine my eyelashes, and then peering further into the pitch-black interior of the barrel, I finally managed to look straight down the microscope tube to be dazzled by an illuminated disc of light. It takes a little while for the eye to train itself to examine what is in the disc. Gajdusek's demonstration to the Fore people was so powerful—after all, the alternatives entirely lacked so concrete a reality—that many accepted his story, even apart from his ability to cure the disease with penicillin. Perhaps some considered the spirochetes in the microscope an amusing example of white-man myth and minor magic, and when another white man arrived querying the origin of disease, they politely returned to him the idea they believed he would be comfortable with. Had Western contact with the Fore people ceased for fifty years, it seems to me entirely possible that a future visitor would discover to his astonishment that the Fore people somehow had knowledge of medical microbiology, despite their largely pretechnological culture.

All three of these stories underline the almost inevitable problems encountered in trying to extract from a "primitive" people their ancient legends. Can you be sure that others have not come before you and destroyed the pristine state of the native myth? Can you be sure that the natives are not humoring you or pulling your leg? Bronislaw Malinowski thought he had discovered a people in the Trobriand Islands who had not worked out the connection between sexual intercourse and childbirth. When asked how children were conceived, they supplied him with an elaborate mythic structure prominently featuring celestial intervention. Amazed, Malinowski objected that was not how it was done at all, and supplied them instead with the version so popular in the West today—including a nine-month gestation period. "Impossible," replied the Melanesians. "Do you not see that woman over there with her six-month-old child? Her husband has been on an extended voyage to another island for two years." Is it more likely that the Melanesians were ignorant of the begetting of children or that they were gently chiding Malinowski? If some peculiar-looking stranger came into my town and asked *me* where babies came from, I'd certainly be tempted to tell him about storks and cabbages. Prescientific people are people. Individually they are as clever as we are. Field interrogation of informants from a different culture is not always easy.

I wonder if the Dogon, having heard from a Westerner an extraordinarily inventive myth about the star Sirius—a star already important in their own mythology—did not carefully play it back to the visiting French anthropologist. Is this not more likely than a visit by extraterrestrial spacefarers to ancient Egypt, with one cluster of hard scientific knowledge, in striking contradiction to common sense, preserved by oral tradition, over the millennia, and only in West Africa?

There are too many loopholes, too many alternative explanations for such a myth to provide reliable evidence of past extraterrestrial contact. If there are extraterrestrials, I think it much more likely that unmanned planetary spacecraft and large radiotelescopes will prove to be the means of their detection.

CHAPTER 7

VENUS AND DR. VELIKOVSKY

◆◆◆

When the movement of the comets is considered and we reflect on the laws of gravity, it will be readily perceived that their approach to the Earth might there cause the most woeful events, bring back the universal deluge, or make it perish in a deluge of fire, shatter it into small dust, or at least turn it from its orbit, drive away its Moon, or, still worse, the Earth itself outside the orbit of Saturn, and inflict upon us a winter several centuries long, which neither men nor animals would be able to bear. The tails even of comets would not be unimportant phenomena, if the comets in taking their departure left them in whole or in part in our atmosphere.

J. H. LAMBERT,
Cosmologische Briefe über die Einrichtung des Weltbaues (1761)

However dangerous might be the shock of a comet, it might be so slight, that it would only do damage at the part of the Earth where it actually struck; perhaps even we might cry quits if while one kingdom were devastated, the rest of the Earth were to enjoy the rarities which a body which came from so far might bring to it. Perhaps we should be very surprised to find that the debris of these masses that we despised were formed of gold and diamonds; but who would be the most astonished, we, or the comet-dwellers, who would be cast on our Earth? What strange beings each would find the other!

MAUPERTUIS,
Lettre sur la comète (1752)

SCIENTISTS, like other human beings, have their hopes and fears, their passions and despondencies—and their strong emotions may sometimes interrupt the course of clear thinking and sound practice. But science is also self-correcting. The most fundamental axioms and conclusions may be challenged. The prevailing hypotheses must survive confrontation with observation. Appeals to authority are impermissible. The steps in a reasoned argument must be set out for all to see. Experiments must be reproducible.

The history of science is full of cases where previously accepted theories and hypotheses have been entirely overthrown, to be replaced by new ideas that more adequately explain the data. While there is an understandable psychological inertia—usually lasting about one generation—such revolutions in scientific thought are widely accepted as a necessary and desirable element of scientific progress. Indeed, the reasoned criticism of a prevailing belief is a service to the proponents of that belief; if they are incapable of defending it, they are well advised to abandon it. This self-questioning and error-correcting aspect of the scientific method is its most striking property, and sets it off from many other areas of human endeavor where credulity is the rule.

The idea of science as a method rather than as a body of knowledge is not widely appreciated outside of science, or indeed in some corridors inside of science. For this reason I and some of my colleagues in the American Association for the Advancement of Science have advocated a regular set of discussions at the annual AAAS meeting of hypotheses that are on the borderlines of science and that have attracted substantial public interest. The idea is not to attempt to settle such issues definitively, but rather to illustrate the process of reasoned disputation, to show how scientists approach a problem that does not lend itself to crisp experimentation, or is unorthodox in its interdisciplinary nature, or otherwise evokes strong emotions.

Vigorous criticism of new ideas is a commonplace in science. While the style of the critique may vary with the character of the critic, overly polite criticism benefits neither the proponents of new ideas nor the scientific enterprise. Any substantive objection is permissible and encouraged; the only exception being that *ad hominem* attacks on the personality or motives of the author are excluded. It does not matter what reason the proponent has for advancing his ideas or what prompts his opponents to criticize them: all that matters is whether the ideas are right or wrong, promising or retrogressive.

For example, here is a summary—of a type that is unusual but not extremely rare—of a paper submitted to the scientific journal *Icarus*, by a qualified referee: "It is the opinion of this reviewer that this paper is absolutely unacceptable for publication in *Icarus*. It is based on no sound scientific research, and at best it is incompetent speculation. The author has not stated his assumptions; the conclusions are unclear, ambiguous and without basis; credit is not given to related work; the figures and tables are unclearly labeled; and the author is obviously unfamiliar with the most basic scientific literature . . ." The referee then goes on to justify his remarks in detail. The paper was rejected for publication. Such rejections are commonly recognized as a boon to science as well as a favor to the author. Most scientists are accustomed to receiving (somewhat milder) referees' criticisms every time they submit a paper to a scientific journal. Almost always the criticisms are helpful. Often a paper revised to take these critiques into account is subsequently accepted for publication. As another example of forthright criticism in the planetary science literature, the interested reader might wish to consult "Comments on *The Jupiter Effect*" by J. Meeus (1975)* and the commentary on it in *Icarus.*

Vigorous criticism is more constructive in science than in some other areas of human endeavor because in science there are adequate standards of validity that can be agreed upon by competent practitioners the world over. The objective of such criticism is not to suppress but rather to encourage the advance of new ideas: those that survive a firm skeptical scrutiny have a fighting chance of being right, or at least useful.

EMOTIONS IN THE scientific community have run very high on the issue of Immanuel Velikovsky's work, especially his first book, *Worlds in Collision,* published in 1950. I know that some scientists were irked because Velikovsky was compared to Einstein, Newton, Darwin and Freud by New York literati and an editor of *Harper's,* but this pique arises from the frailty of human nature rather than the judgment of the scientist. The two together often inhabit the same individual. Others were dismayed at the use of Indian, Chinese, Aztec, Assyrian or Biblical texts to argue for extremely heterodox views in celestial mechanics. Also, I suspect, not many physicists or celestial mechanicians are comfortably fluent in such languages or are familiar with such texts.

My own view is that no matter how unorthodox the reasoning process or how unpalatable the conclusions, there is no excuse for any attempt to suppress new ideas—least of all by scientists. Therefore I was very pleased that the AAAS held a discussion on *Worlds in Collision,* in which Velikovsky took part.

In reading the critical literature in advance, I was surprised at how little of it there is and how rarely it approaches the central points of Velikovsky's thesis. In fact, neither the critics nor the proponents of Velikovsky seem to have read him carefully, and I even seem to find some cases where Velikovsky has not read

* Citations to references in this chapter are given at the end of the book.

Velikovsky carefully. Perhaps the publication of most of the AAAS symposium (Goldsmith, 1977) as well as the present chapter, the principal conclusions of which were presented at the symposium, will help to clarify the issues.

In this chapter I have done my best to analyze critically the thesis of *Worlds in Collision,* to approach the problem both on Velikovsky's terms and on mine—that is, to keep firmly in mind the ancient writings that are the focus of his argument, but at the same time to confront his conclusions with the facts and the logic I have at my command.

Velikovsky's principal thesis is that major events in the history of both the Earth and the other planets in the solar system have been dominated by catastrophism rather than by uniformitarianism. These are fancy words used by geologists to summarize a major debate they had during the infancy of their science which apparently culminated, between 1785 and 1830, in the work of James Hutton and Charles Lyell, in favor of the uniformitarians. Both the names and the practices of these two sects evoke familiar theological antecedents. A uniformitarian holds that landforms on Earth have been produced by processes we can observe to be operating today, provided they operate over immense vistas of time. A catastrophist holds that a small number of violent events, occupying much shorter periods of time, are adequate. Catastrophism began largely in the minds of those geologists who accepted a literal interpretation of the Book of Genesis, and in particular the account of the Noahic flood. It is clearly no use arguing against the catastrophist viewpoint to say that we have never seen such a catastrophe in our lifetimes. The hypothesis requires only rare events. But if we can show that there is adequate time for processes we can all observe operating today to produce the landform or event in question, then there is at least no necessity for the catastrophist hypothesis. Obviously both uniformitarian and catastrophic processes can have been at work—and almost certainly both were—in the history of our planet.

Velikovsky holds that in the relatively recent history of the Earth there has been a set of celestial catastrophes, near-collisions with comets, small planets and large planets. There is nothing absurd in the possibility of cosmic collisions. Astronomers in the past have not hesitated to invoke collisions to explain natural phenomena. For example, Spitzer and Baade (1951) proposed that extragalactic radio sources may be produced by the collisions of whole galaxies, containing hundreds of billions of stars. This thesis has now been abandoned, not because cosmic collisions are unthinkable, but because the frequency and properties of such collisions do not match what is now known about such radio sources. A still popular theory of the energy source of quasars is multiple stellar collisions at the centers of galaxies—where, in any case, catastrophic events must be common.

Collisions and catastrophism are part and parcel of modern astronomy, and have been for many centuries (see the epigraphs at the beginning of this chapter). For example, in the early history of the solar system, when there were probably many more objects about than there are now—including objects on very eccentric orbits—collisions may have been frequent. Lecar and Franklin (1973) investigate hundreds of collisions occurring in a period of only a few thousand years in the early history of the asteroid belt, to understand the present configuration of this region of the solar system. In another paper, entitled "Cometary Collisions and Geological Periods," Harold Urey (1973) investigates a range of consequences, including the production of earthquakes and the heating of the oceans, which might attend the collision with the Earth of a comet of average mass of about 10^{18} grams. The Tunguska event of 1908, in which a Siberian forest was leveled, is often attributed to the collision with the Earth of a small comet. The cratered surfaces of Mercury, Mars, Phobos, Deimos and the Moon bear eloquent testimony to the fact that there have been abundant collisions during the history of the solar system. There is nothing unorthodox about the idea of cosmic catastrophes, and this is a view that has been common in solar system physics at least back to the late-nineteenth-century studies of the lunar surface by G. K. Gilbert, the first director of the U.S. Geological Survey.

What, then, is all the furor about? It is about the time scale and the adequacy of the purported evidence. In the 4.6 billion-year history of the solar system, many collisions must have occurred. But have there been major collisions in the last 3,500 years, and can the study of ancient writings demonstrate such collisions? That is the nub of the issue.

VELIKOVSKY has called attention to a wide range of stories and legends, held by diverse peoples, separated by great distances, which stories show remarkable similarities and concordances. I am not expert in the cultures or languages of any of these peoples, but I find the concatenation of legends Velikovsky has accumulated stunning. It is true that some experts in these cultures are less impressed. I can remember vividly discussing *Worlds in Collision* with a distinguished professor of Semitics at a leading university. He said something like "The Assyriology, Egyptology, Biblical scholarship and all of that Talmudic and Midrashic *pilpul* is, of course, nonsense; but I was impressed by the astronomy." I had rather the opposite view. But let me not be swayed by the opinions of others. My own position is that if even 20 percent of the legendary concordances that Velikovsky produces are real, there is something important to be explained. Furthermore, there is an impressive array of cases in the history of archaeology—from Heinrich Schliemann at Troy to Yigael Yadin at Masada—where the descriptions in ancient writings have subsequently been validated as fact.

Now, if a variety of widely separated cultures share what is palpably the same legend, how can this be understood? There seem to be four possibilities: common observation, diffusion, brain wiring and coincidence. Let us consider these in turn.

Common Observation: One explanation is that the

cultures in question all witnessed a common event and interpreted it in the same way. There may, of course, be more than one view of what this common event was.

Diffusion: The legend originated within one culture only, but during the frequent and distant migrations of mankind, gradually spread with some changes among many apparently diverse cultures. A trivial example is the Santa Claus legend in America which evolved from the European Saint Nicholas (Claus is short for Nicholas in German), the patron saint of children, and which ultimately is derived from pre-Christian tradition.

Brain Wiring: A hypothesis sometimes also known as racial memory or the collective unconscious. It holds that there are certain ideas, archetypes, legendary figures, and stories that are intrinsic to human beings at birth, perhaps in the same way that a newborn baboon knows to fear a snake, and a bird raised in isolation from other birds knows how to build a nest. It is apparent that if a tale derived from observation or from diffusion resonated with the "brain wiring," it is more likely to be culturally retained.

Coincidence: Purely by chance two independently derived legends may have similar content. In practice, this hypothesis fades into the brain-wiring hypothesis.

IF WE ARE TO ASSESS critically such apparent concordances, there are some obvious precautions that must first be taken. Do the stories really say the same thing or have the same essential elements? If they are interpreted as due to common observations, do they date from the same period? Can we exclude the possibility of physical contact between representatives of the cultures in question in or before the epoch under discussion? Velikovsky is clearly opting for the common-observation hypothesis, but he seems to dismiss the diffusion hypothesis far too casually; for example, he says (page 303*): "How could unusual motifs of folklore reach isolated islands, where the aborigines do not have any means of crossing the sea?" I am not sure

* The page numbers refer to the canonical English-language edition (Velikovsky, 1950).

which islands and which aborigines Velikovsky refers to here, but it is apparent that the inhabitants of an island had to have gotten there somehow. I do not think that Velikovsky believes in a separate creation in the Gilbert and Ellice Islands, say. For Polynesia and Melanesia there is now extensive evidence of abundant sea voyages of lengths of many thousands of kilometers within the last millennium, and probably much earlier (Dodd, 1972).

Or how, for example, would Velikovsky explain the fact that the Toltec name for "god" seems to have been *teo,* as in the great pyramid city of Teotihuacán ("City of the Gods") near present-day Mexico City, where it is called San Juan Teotihuacán? There is no common celestial event that could conceivably explain this concordance. Toltec and Nahuatl are non-Indoeuropean languages, and it seems unlikely that the *word* for "god" would be wired into all human brains. Yet *teo* is a clear cognate of the common Indoeuropean root for "god," preserved, among other places, in the words "deity" and "theology." The preferred hypotheses in this case are coincidence or diffusion. There is some evidence for pre-Columbian contact between the Old and New Worlds. But coincidence is also not to be taken lightly: if we compare two languages, each with tens of thousands of words, spoken by human beings with identical larynxes, tongues and teeth, it should not be surprising if a few words are coincidentally identical. Likewise, we should not be surprised if a few elements of a few legends are coincidentally identical. But I believe that *all* of the concordances Velikovsky produces can be explained away in this manner.

Let us take an example of Velikovsky's approach to this question. He points to certain concordant stories, directly or vaguely connected with celestial events, that refer to a witch, a mouse, a scorpion or a dragon (pages 77, 264, 305, 306, 310). His explanation: divers comets, upon close approach to the Earth, were tidally or electrically distorted and gave the form of a witch, a mouse, a scorpion or a dragon, clearly interpretable as the same animal to culturally isolated peoples of very different backgrounds. No attempt is made to show that such a clear form—for example, a woman riding a broomstick and topped by a pointed hat—could have been produced in this way, even if we grant the hypothesis of a close approach to the Earth by a comet. Our experience with Rorschach and other psychological projective tests is that different people see the same nonrepresentational image in different ways. Velikovsky even goes so far as to believe that a close approach to the Earth by "a star" he evidently identifies with the planet Mars so distorted it that it took on the clear shape (page 264) of lions, jackals, dogs, pigs and fish; and in his opinion this explains the worship of animals by the Egyptians. This is not very impressive reasoning. We might just as well assume that the whole menagerie was capable of independent flight in the second millennium B.C. and be done with it. A much more likely hypothesis is diffusion. Indeed, I have in a different context spent a fair amount of time studying the dragon legends on the planet Earth, and I am impressed at how different these mythical beasts, all called dragons by Western writers, really are.

As another example, consider the argument of Chapter 8, Part 2 of *Worlds in Collision.* Velikovsky claims a world-wide tendency in ancient cultures to believe at various times that the year has 360 days, that the month has thirty-six days, and that the year has ten months. Velikovsky offers no justification in physics for this, but argues that ancient astronomers could hardly have been so poor at their trade as to slip five days each year or six days each lunation. Fairly soon the night would be brilliant with moonlight at the astrologically official new moon, snowstorms would be falling in July, and the astrologers would be hung by their ears. Having had some experience with modern astronomers, I am not as confident as Velikovsky is in the unerring computational precision of ancient astronomers. Velikovsky proposes that these aberrant calendrical conventions reflect real changes in the length of the day, month and/or year—and that they are evidence of close approaches to the Earth-Moon system by comets, planets and other celestial visitors.

There is an alternative explanation, which derives from the fact that there are not a whole number of lunations in a solar year, nor a whole number of days in a lunation. These incommensurabilities will be galling to a culture that has recently invented arithmetic but has not yet gotten as far as large numbers or fractions. As an inconvenience, these incommensurabilities are felt even today by religious Muslims and Jews who discover that Ramadan and Passover, respectively, occur from year to year on rather different days of the solar calendar. There is a clear whole-number chauvinism in human affairs, most easily discerned in discussing arithmetic with four-year-olds; and this seems to be a much more plausible explanation of these calendrical irregularities, if they existed.

Three hundred and sixty days a year provides an obvious (temporary) convenience for a civilization with base-60 arithmetic, as the Sumerian, Akkadian, Assyrian and Babylonian cultures. Likewise, thirty days per month or ten months per year might be attractive to enthusiasts of base-10 arithmetic. I wonder if we do not see here an echo of the collision between chauvinists of base-60 arithmetic and chauvinists of base-10 arithmetic, rather than a collision of Mars with Earth. It is true that the tribe of ancient astrologers may have been dramatically depleted as the various calendars rapidly slipped out of phase, but that was an occupational hazard, and at least it removed the mental agony of dealing with fractions. In fact, sloppy quantitative thinking appears to be the hallmark of this whole subject.

An expert on early time-reckoning (Leach, 1957) points out that in ancient cultures the first eight or ten months of the year are named, but the last few months, because of their economic unimportance in an agricultural society, are not. Our month December, named after the Latin *decem,* means the tenth, not the twelfth, month. (September = seventh, October = eighth, November = ninth, as well.) Because of the large numbers involved, prescientific peoples characteristically do not count days of the year, although they are assiduous in counting months. A leading historian of ancient science and mathematics, Otto Neugebauer (1957), remarks that, both in Mesopotamia and in Egypt, two separate and mutually exclusive calendars were maintained: a civil calendar whose hallmark was computational convenience, and a frequently updated agricultural calendar—messier to deal with, but much closer to the seasonal and astronomical realities. Many ancient cultures solved the two-calendar problem by simply adding a five-day holiday on at the end of the year. I hardly think that the existence of 360-day years in the calendrical conventions of prescientific peoples is compelling evidence that then there really were 360 rather than 365¼ rotations in one revolution of Earth about the Sun.

This question can, in principle, be resolved by examining coral growth rings, which are now known to show with some accuracy the number of days per month and the number of days per year, the former only for intertidal corals. There appears to be no sign of major excursions in recent times from the present number of days in a lunation or a year, and the gradual shortening (not lengthening) of the day and the month with respect to the year as we go back in time is found to be consistent with tidal theory and the conservation of energy and angular momentum within the Earth-Moon system, without appeal to cometary or other exogenous intervention.

Another problem with Velikovsky's method is the suspicion that vaguely similar stories may refer to quite different periods. This question of the synchronism of legends is almost entirely ignored in *Worlds in Collision,* although it is treated in some of Velikovsky's later works. For example (page 31), Velikovsky notes that the idea of four ancient ages terminated by catastrophe is common to Indian as well as to Western sacred writing. However, in the *Bhagavad Gita* and in the Vedas, widely divergent numbers of such ages, including an infinity of them, are given; but, more interesting, the duration of the ages between major catastrophes is specified (see, for example, Campbell, 1974) as billions of years. This does not match very well Velikovsky's chronology, which requires hundreds or thousands of years. Here Velikovsky's hypothesis and the data that purport to support it differ by a factor of about a million. Or (page 91) vaguely similar discussions of vulcanism and lava flows in Greek, Mexican and Biblical traditions are quoted. There is no attempt made to show that they refer to even approximately comparable times and, since lava has flowed in historical times in all three areas, no common exogenous event is necessary to interpret such stories.

Despite copious references, there also seem to me to be a large number of critical and undemonstrated assumptions in Velikovsky's argument. Let me mention just a few of them. There is the very interesting idea that any mythological references by any people to any god that also corresponds to a celestial body represents in fact a direct observation of that celestial body. It is a daring hypothesis, although I am not sure what one is to do with Jupiter appearing as a swan to Leda, and as a shower of gold to Danaë. On page 247 the hypothesis that gods and planets are identical is used to date the time of Homer. In any case, when Hesiod and Homer refer to Athena being born full-grown from the head of Zeus, Velikovsky takes Hesiod and Homer at their word and assumes that the celestial body Athena was ejected by the planet Jupiter. But what *is* the celestial body Athena? Repeatedly it is identified with the planet Venus (Part 1, Chapter 9, and many other places in the text). One would scarcely guess from reading *Worlds in Collision* that the Greeks characteristically identified Aphrodite with Venus, and Athena with no celestial body whatever. What is more, Athena and Aphrodite were "contemporaneous" goddesses, both being born at the time Zeus was king of the gods. On page 251 Velikovsky notes that Lucian "is unaware that Athene is the goddess of the planet Venus." Poor Lucian seems to be under the misconception that Aphrodite is the goddess of the planet Venus. But in the footnote on page 361 there appears to be a slip, and here Velikovsky for the first and only time uses the form "Venus (Aphrodite)." On page 247 we hear of Aphrodite, the goddess of the Moon. Who, then, was Artemis, the sister of Apollo the Sun, or, earlier, Selene? There may be good justification, for all I know, in identifying Athena with Venus, but it is far from the prevailing wisdom either now or two thousand years ago, and it is central to Velikovsky's argument. It does not increase our confidence in the presentation of less familiar myths when the celestial identification of Athena is glossed over so lightly.

Other critical statements which are given extremely inadequate justification, and which are central to one or more of Velikovsky's major themes, are: the statement (page 283) that "Meteorites, when entering the earth's atmosphere, make a frightful din," when they are generally observed to be silent; the statement (page 114) that "a thunderbolt, when striking a magnet, reverses the poles of the magnet"; the translation (page 51) of "Barad" as meteorites; and the contention (page 85) "as is known, Pallas was another name for Typhon." On page 179 a principle is implied that when two gods are hyphenated in a joint name, it indicates an attribute of a celestial body—as, for example, Ashteroth-Karnaim, a horned Venus, which Velikovsky interprets as a crescent Venus and evidence that Venus was once close enough to Earth to have its phases discernible to the naked eye. But what does this principle imply, for example, for the god Ammon-Ra? Did the Egyptians see the sun (Ra) as a ram (Ammon)?

There is a contention (page 63) that instead of the tenth plague of Exodus killing the "first born" of Egypt, what is intended is the killing of the "chosen." This is a rather serious matter and at least raises the suspicion that where the Bible is inconsistent with Velikovsky's hypothesis, Velikovsky retranslates the Bible. The foregoing queries may all have simple answers, but the answers are not to be found easily in *Worlds in Collision.*

I do not mean to suggest that all of Velikovsky's legendary concordances and ancient scholarship are similarly flawed, but many of them seem to be, and the remainder may well have alternative, for example diffusionist, origins.

With the situation in legend and myth as fuzzy as this, any corroboratory evidence from other sources would be welcomed by those who support Velikovsky's argument. I am struck by the absence of any confirming evidence in art. There is a wide range of paintings, bas-reliefs, cylinder seals and other *objets d'art* produced by humanity and going back to at least 10,000 B.C. They represent all of the subjects, especially mythological subjects, important to the cultures that created them. Astronomical events are not uncommon in such works of art. Recently (Brandt, *et al.*, 1974), impressive evidence has been uncovered in cave paintings in the American Southwest of contemporary observations of the Crab Supernova event of the year 1054, which was also recorded in Chinese, Japanese and Arab annals. Appeals have been made to archaeologists for information on cave painting representations of the earlier Gum Supernova (Brandt, *et al.*, 1971). But supernova events are not nearly so impressive as the close approach of another planet with attendant interplanetary tendrils and lightning discharges connecting it to Earth. There are many unflooded caves at high altitudes, distant from the sea. If the Velikovskian catastrophes occurred, why are there no contemporary graphic records of them?

I therefore cannot find the legendary base of Velikovsky's hypothesis at all compelling. If, nevertheless, his notion of recent planetary collisions and global catastrophism were strongly supported by physical evidence, we might be tempted to give it some credence. If the physical evidence is not, however, very strong, the mythological evidence will surely not stand by itself.

LET ME GIVE a short summary of my understanding of the basic features of Velikovsky's principal hypothesis. I will relate it to the events described in the Book of *Exodus,* although the stories of many other cultures are said to be consistent with the events described in *Exodus:*

The planet Jupiter disgorged a large comet, which made a grazing collision with Earth around 1500 B.C. The various plagues and Pharaonic tribulations of the Book of *Exodus* all derive directly or indirectly from this cometary encounter. Material which made the river Nile turn to blood drops from the comet. The vermin described in *Exodus* are produced by the comet—flies and perhaps scarabs drop out of the comet, while indigenous terrestrial frogs are induced by the heat of the comet to multiply. Earthquakes produced by the comet level Egyptian but not Hebrew dwellings. (The only thing that does not seem to drop from the comet is cholesterol to harden Pharaoh's heart.)

All this evidently falls from the coma of the comet, because at the moment that Moses lifts his rod and stretches out his hand, the "Red Sea" parts—due either to the gravitational tidal field of the comet or to some unspecified electrical or magnetic interaction between the comet and the "Red Sea." Then, when the Hebrews have successfully crossed, the comet has evidently passed sufficiently farther on for the parted waters to flow back and drown the host of Pharaoh. The Children of Israel during their subsequent forty years of wandering in the Wilderness of Sin are nourished by manna from heaven, which turns out to be hydrocarbons (or carbohydrates) from the tail of the comet.

Another reading of *Worlds in Collision* makes it appear that the plagues and the Red Sea events represent two different passages of the comet, separated by a month or two. Then after the death of Moses and the passing of the mantle of leadership to Joshua, the same comet comes screeching back for another grazing collision with the Earth. At the moment that Joshua says "Sun, stand thou still upon Gibeon; and thou, Moon, in the valley of Ajalon," the Earth—perhaps because of tidal interaction again, or perhaps because of an unspecified magnetic induction in the crust of the Earth—obligingly ceases its rotation, to permit Joshua victory in battle. The comet then makes a near-collision with Mars, so violent as to eject it out of its orbit so it makes two near-collisions with the Earth which destroy the army of Sennacherib, the Assyrian king, as he was making life miserable for some subsequent generation of Israelites. The net result was to eject Mars into its present orbit and the comet into a circular orbit around the Sun, where it became the planet Venus—which previously, Velikovsky believes, did not exist. The Earth meantime had somehow begun rotating again at almost exactly the same rate as before these encounters. No subsequent aberrant planetary behavior has occurred since about the seventh century B.C., although it might have been common in the Second Millennium.

That this is a remarkable story no one—proponents and opponents alike—will disagree. Whether it is a likely story is, fortunately, amenable to scientific inquiry. Velikovsky's hypothesis makes certain predictions and deductions: that comets are ejected from planets; that comets are likely to make near or grazing collisions with planets; that vermin live in comets and in the atmospheres of Jupiter and Venus; that carbohydrates can be found in the same places; that enough carbohydrates fell in the Sinai peninsula for nourishment during forty years of wandering in the desert; that eccentric cometary or planetary orbits can be circularized in a period of hundreds of years; that volcanic and tectonic events on Earth and impact events on the Moon were contemporaneous with these catastrophes; and so on. I will discuss each of these ideas, as well as some others—for example, that the surface of Venus is hot, which is clearly less central to his hypothesis, but which has been widely advertised as powerful *post hoc* support of it. I will also examine an occasional additional "prediction" of Velikovsky—for example, that the Martian polar caps are carbon or carbohydrates. My conclusion is that when Velikovsky is original he is very likely wrong, and that when he is right the idea has been pre-empted by earlier workers. There are also a large number of cases where he is neither right nor original. The question of originality is important because of circumstances—for example, the high surface temperature of Venus—which are said to have been predicted by Velikovsky at a time when everyone else was imagining something very different. As we shall see, this is not quite the case.

In the following discussion, I will try to use simple quantitative reasoning as much as possible. Quantitative arguments are obviously a finer mesh with which to sift hypotheses than qualitative arguments. For example, if I say that a large tidal wave engulfed the Earth, there is a wide range of catastrophes—from the flooding of littoral regions to global inundation—which might be pointed to as support for my contention. But if I specify a tide 100 miles high, I must be talking about the latter, and moreover, there might be some critical evidence to counterindicate or support a tide of such dimensions. However, so as to make the quantitative arguments tractable to the reader who is not very familiar with elementary physics, I have tried, particularly in the Appendices (following the References), to state all the essential steps in the quantitative development, using the simplest arguments that preserve the essential physics. Perhaps I need not mention that such quantitative testing of hypotheses is entirely routine in the physical and biological sciences today. By rejecting the hypotheses that do not meet these standards of analysis, we are able to move swiftly to hypotheses in better concordance with the facts.

There is one further point about scientific method that must be made. Not all scientific statements have equal weight. Newtonian dynamics and the laws of conservation of energy and angular momentum are on extremely firm footing. Literally millions of separate experiments have been performed on their validity—not just on Earth, but, using the observational techniques of modern astrophysics, elsewhere in the solar system, in other star systems, and even in other galaxies. On the other hand, questions on the nature of planetary surfaces, atmospheres and interiors are on much weaker footing, as the substantial debates on these matters by planetary scientists in recent years clearly indicate. A good example of this distinction is the appearance in

1975 of Comet Kohoutek. This comet had first been observed at a great distance from the Sun. On the basis of the early observations, two predictions were made. The first concerned the orbit of Comet Kohoutek—where it would be found at future times, when it would be observable from the Earth before sunrise, when after sunset—predictions based on Newtonian dynamics. These predictions were correct to within a gnat's eyelash. The second prediction concerned the brightness of the comet. This was based on the guessed rate of vaporization of cometary ices to make a large cometary tail which brightly reflects sunlight. This prediction was painfully in error, and the comet—far from rivaling Venus in brightness—could not be seen at all by most naked-eye observers. But vaporization rates depend on the detailed chemistry and geometrical form of the comet, which we know poorly at best. The same distinction between well-founded scientific arguments, and arguments based on a physics or chemistry that we do not fully understand, must be borne in mind in any analysis of *Worlds in Collision*. Arguments based on Newtonian dynamics or the conservation laws of physics must be given very great weight. Arguments based on planetary surface properties, for example, must have correspondingly lesser weights. We will find that Velikovsky's arguments run into extremely grave difficulties on both these scores, but the one set of difficulties is far more damaging than the other.

PROBLEM I

THE EJECTION OF VENUS BY JUPITER

VELIKOVSKY's hypothesis begins with an event that has never been observed by astronomers and that is inconsistent with much that we know about planetary and cometary physics, namely, the ejection of an object of planetary dimensions from Jupiter, perhaps by its collision with some other giant planet. Such a propagation of catastrophes, Velikovsky promised, would be "the theme of the sequel to *Worlds in Collision*" (page 373). Thirty years later, no sequel of this description has appeared. From the fact that the aphelia (the greatest distances from the Sun) of the orbits of short-period comets have a statistical tendency to lie near Jupiter, Laplace and other early astronomers hypothesized that Jupiter was the source of such comets. This is an unnecessary hypothesis because we now know that long-period comets may be transferred to short-period trajectories by the perturbations of Jupiter; this view has not been advocated for a century or two except by the Soviet astronomer V. S. Vsekhsviatsky, who seems to believe that the moons of Jupiter eject comets out of giant volcanoes.

To escape from Jupiter, such a comet must have a kinetic energy of $\frac{1}{2} mv_e^2$, where m is the cometary mass and v_e is the escape velocity from Jupiter, which is about 60 km/sec. Whatever the ejection event—volcanoes or collisions—some significant fraction, at least 10 percent, of this kinetic energy will go into heating the comet. The minimum kinetic energy per unit mass ejected is then $\frac{1}{2} v_e^2 = 1.3 \times 10^{13}$ ergs per gram, and the quantity that goes into heating is more than 2.5×10^{12} erg/gram. The latent heat of fusion of rock is about 4×10^9 ergs per gram. This is the heat that must be applied to convert hot solid rock near the melting point into a fluid lava. About 10^{11} ergs/gm must be applied to raise rocks at low temperatures to their melting point. Thus, any event that ejected a comet or a planet from Jupiter would have brought it to a temperature of at least several thousands of degrees, and whether composed of rocks, ices or organic compounds, would have completely melted it. It is even possible that it would have been entirely reduced to a rain of self-gravitating small dust particles and atoms, which does not describe the planet Venus particularly well. (Incidentally, this would appear to be a good Velikovskian argument for the high temperature of the surface of Venus, but, as described below, this is not his argument.)

Another problem is that the escape velocity from the

Sun's gravity at the distance of Jupiter is about 20 km/sec. The ejection mechanism from Jupiter does not, of course, know this. Thus, if the comet leaves Jupiter at velocities less than about 60 km/sec, the comet will fall back to Jupiter; if greater than about $[(20)^2 + (60)^2]^{1/2} = 63$ km/sec, it will escape from the solar system. There is only a narrow and therefore unlikely range of velocities consistent with Velikovsky's hypothesis.

A further problem is that the mass of Venus is very large—more than 5×10^{27} grams, or possibly larger originally, on Velikovsky's hypothesis, before it passed close to the Sun. The total kinetic energy required to propel Venus to Jovian escape velocity is then easily calculated to be on the order of 10^{41} ergs, which is equivalent to all the energy radiated by the Sun to space in an entire year, and one hundred million times more powerful than the largest solar flare ever observed. We are asked to believe, without any further evidence or discussion, an ejection event vastly more powerful than anything on the Sun, which is a far more energetic object than Jupiter.

Any process that makes large objects makes more small objects. This is especially true in a situation dominated by collisions, as in Velikovsky's hypothesis. Here the comminution physics is well known and a particle one-tenth as large as our biggest particle should be a hundred or a thousand times more abundant. Indeed, Velikovsky has stones falling from the skies in the wake of his hypothesized planetary encounters, and imagines Venus and Mars trailing swarms of boulders; the Mars swarm, he says, led to the destruction of the armies of Sennacherib. But if this is true, if we had near-collisions with objects of planetary mass only thousands of years ago, we should have been bombarded by objects of lunar mass hundreds of years ago; and bombardment by objects that can make craters a mile or so across should be happening every second Tuesday. Yet there is no sign, on either the Earth or the Moon, of frequent recent collisions with such lower mass objects. Instead, the few objects that, as a steady-state population, are moving in orbits that might collide with the Moon are just adequate, over geological time, to explain the number of craters observed on the lunar maria. The absence of a great many small objects with orbits crossing the orbit of the Earth is another fundamental objection to Velikovsky's basic thesis.

PROBLEM II

REPEATED COLLISIONS AMONG THE EARTH, VENUS AND MARS

"THAT A COMET may strike our planet is not very probable, but the idea is not absurd" (page 40.) This is precisely correct: it remains only to calculate the probabilities, which Velikovsky has unfortunately left undone.

Fortunately, the relevant physics is extremely simple and can be performed to order of magnitude even without any consideration of gravitation. Objects on highly eccentric orbits, traveling from the vicinity of Jupiter to the vicinity of the Earth, are traveling at such high speeds that their mutual gravitational attraction to the object with which they are about to have a grazing collision plays a negligible role in determining the trajectory. The calculation is performed in Appendix 1, where we see that a single "comet" with aphelion (far point from the Sun) near the orbit of Jupiter and perihelion (near point to the Sun) inside the orbit of Venus should take at least 30 million years before it impacts the Earth. We also find in Appendix 1 that if the object is a member of the currently observed family of objects on such trajectories, the lifetime against collision exceeds the age of the solar system.

But let us take the number 30 million years to give the maximum quantitative bias in favor of Velikovsky. Therefore, the odds against a collision with the Earth in any given year is 3×10^7 to 1; the odds against it in any given millennium are 30,000 to 1. But Velikovsky has (see, e.g., page 388) not one but *five* or *six* near-collisions among Venus, Mars and the Earth—all of which seem to be statistically independent events; that is, by his own account, there does not seem to be a regular set of grazing collisions determined by the relative orbital periods of the three planets. (If there were, we would have to ask the probability that so remarkable a play in the game of planetary billiards could arise within Velikovsky's time constraints.) If the probabilities are independent, then the joint probability of five such encounters in the same millennium is on the short side of $(3 \times 10^7/10^3)^{-5} = (3 \times 10^4)^{-5} = 4.1 \times 10^{-23}$, or almost 100 billion trillion to 1 odds. For six encounters in the same millennium the odds rise to $(3 \times 10^7/10^3)^{-6} = (3 \times 10^4)^{-6} = 7.3 \times 10^{-28}$, or about a trillion quadrillion to 1 odds. Actually, these are lower limits—both for the reason given above and because close encounters with Jupiter are likely to eject the impacting object out of the solar system altogether, rather as Jupiter ejected the Pioneer 10 spacecraft. These odds are a proper calibration of the validity of Velikovsky's hypothesis, even if there were no other difficulties with it. Hypotheses with such small odds in their favor are usually said to be untenable. With the other problems mentioned both above and below, the probability that the full thesis of *Worlds in Collision* is correct becomes negligible.

PROBLEM III
THE EARTH'S ROTATION

MUCH OF THE indignation directed toward *Worlds in Collision* seems to have arisen from Velikovsky's interpretation of the story of Joshua and related legends as implying that the Earth's rotation was once braked to a halt. The image that the most outraged protesters seem to have had in mind is that in the movie version of H. G. Wells's story "The Man Who Could Work Miracles": The Earth is miraculously stopped from rotating but, through an oversight, no provision is made for objects that are not nailed down, which then continue moving at their usual rate and therefore fly off the Earth at a speed of 1,000 miles per hour. But it is easy to see (Appendix 2) that a gradual deceleration of the Earth's rotation at 10^{-2}g or so could occur in a period of much less than a day. Then no one would fly off, and even stalactites and other delicate geomorphological forms could survive. Likewise, we see in Appendix 2 that the energy required to brake the Earth is not enough to melt it, although it would result in a noticeable increase in temperature: the oceans would have been raised to the boiling point of water, an event that seems to have been overlooked by Velikovsky's ancient sources.

These are, however, not the most serious objections to Velikovsky's exegesis of Joshua. Perhaps the most serious is at the other end: How does the Earth get started up again, rotating at approximately the same rate of spin? The Earth cannot do it by itself, because of the law of the conservation of angular momentum. Velikovsky does not even seem to be aware that this is a problem.

Nor is there any hint that braking the Earth to a "halt" by cometary collision is any less likely than any other resulting spin. In fact, the chance of precisely canceling the Earth's rotational angular momentum in a cometary encounter is tiny; and the probability that subsequent encounters, were they to occur, would start the Earth spinning again even approximately once every twenty-four hours is tiny squared.

Velikovsky is vague about the mechanism that is supposed to have braked the Earth's rotation. Perhaps it is tidal gravitational; perhaps it is magnetic. Both of these fields produce forces that decline very rapidly with distance. While gravity declines as the inverse square of the distance, tides decline as the inverse cube, and the tidal couple as the inverse sixth power. The magnetic dipole field declines as the inverse cube and any equivalent magnetic tides fall off even more steeply than gravitational tides. Therefore, the braking effect is almost entirely at the distance of closest approach. The characteristic time of this closest approach is clearly about 2R/v, where R is the radius of the Earth and v the relative velocity of the comet and the Earth. With v about 25 km/sec, the characteristic time works out to be under ten minutes. This is the full time available for the total effect of the comet on the rotation of the Earth. The corresponding acceleration is less than 0.1 g, so armies still do not fly off into space. But the characteristic time for acoustic propagation within the Earth—the minimum time for an exterior influence to make itself felt on the Earth as a whole—is eighty-five minutes. Thus, no cometary influence even in grazing collision could make the Sun stand still upon Gibeon.

Velikovsky's account of the history of the Earth's rotation is difficult to follow. On page 236 we have an account of the motion of the Sun in the sky which by accident conforms to the appearance and apparent motion of the Sun as seen from the surface of Mercury, but not from the surface of the Earth; and on page 385 we seem to have an aperture to a wholesale retreat by Velikovsky—for here he suggests that what happened was not any change in the angular velocity of rotation of the Earth, but rather a motion in the course of few hours of the angular momentum vector of the Earth from pointing approximately at right angles to the ecliptic plane as it does today to pointing in the direction of the Sun, like the planet Uranus. Quite apart from extremely grave problems in the physics of this suggestion, it is inconsistent with Velikovsky's own argument, because earlier he has laid great weight on the fact that Eurasian and Near Eastern cultures reported prolonged day, while North American cultures reported prolonged night. In this variant there would be no explanation of the reports from Mexico. I think I see in this instance Velikovsky hedging on or forgetting his own strongest arguments from ancient writings. On page 386 we have a qualitative argument, not reproduced, claiming that the Earth could have been braked to a halt by a strong magnetic field. The field strength required is not mentioned but would clearly (cf. calculations in Appendix 4) have to be enormous. There is no sign in rock magnetization of terrestrial rocks ever having been subjected to such strong field strengths and, what is equally important, we have quite firm evidence from both Soviet and American spacecraft that the magnetic-field strength of Venus is negligibly small—far less than the Earth's own surface field of 0.5 gauss, which would itself have been inadequate for Velikovsky's purpose.

PROBLEM IV

TERRESTRIAL GEOLOGY AND LUNAR CRATERS

REASONABLY enough, Velikovsky believes that a near-collision of another planet with the Earth might have had dramatic consequences here—by gravitational tidal, electrical or magnetic influences (Velikovsky is not very clear on this). He believes (pages 96 and 97) "that in the days of the Exodus, when the world was shaken and rocked . . . *all* volcanoes vomited lava and *all* continents quaked." (My emphasis.)

There seems little doubt that earthquakes would have accompanied such a near-collision. Apollo lunar seismometers have found that moonquakes are most common during lunar perigee, when the Earth is closest to the Moon, and there are at least some hints of earthquakes at the same time. But the claim that there were extensive lava flows and volcanism involving "all volcanoes" is quite another story. Volcanic lavas are easily dated, and what Velikovsky should produce is a histogram of the number of lava flows on Earth as a function of time. Such a histogram will, I believe, show that not all volcanoes were active between 1500 and 600 B.C., and that there is nothing particularly remarkable about the volcanism of that epoch.

Velikovsky believes (page 115) that reversals of the geomagnetic field are produced by cometary close approaches. Yet the record from rock magnetization is clear—such reversals occur about every million years, and not in the last few thousand, and they recur more or less like clockwork. Is there a clock in Jupiter that aims comets at the Earth every million years? The conventional view is that the Earth experiences a polarity reversal of the self-sustaining dynamo that produces the Earth's magnetic field; it seems a much more likely explanation.

Velikovsky's contention that mountain building occurred a few thousand years ago is belied by all the geological evidence, which puts those times at tens of millions of years ago and earlier. The idea that mammoths were deep-frozen by a rapid movement of the Earth's geographical pole a few thousands of years ago can be tested—for example, by carbon-14 or amino-acid racemization dating. I should be very surprised if a very recent age results from such tests.

Velikovsky believes that the Moon, not immune to the catastrophes which befell the Earth, had similar tectonic events occur on its surface a few thousand years ago, and that many of its craters were formed then (see Part 2, Chapter 9). There are some problems with this idea as well: samples returned from the Moon in the Apollo missions show no rocks melted more recently than a few hundred million years ago.

Furthermore, if lunar craters were to have formed abundantly 2,700 or 3,500 years ago, there must have been a similar production at the same time of terrestrial craters larger than a kilometer across. Erosion on the Earth's surface is inadequate to remove any crater of this size in 2,700 years. There are not large numbers of terrestrial craters of this size and age; indeed, there is not a single one. On these questions Velikovsky seems to have ignored the critical evidence. When the evidence is examined, it strongly counterindicates his hypothesis.

Velikovsky believes that the close passage of Venus or Mars to the Earth would have produced tides at least miles high (pages 70 and 71); in fact, if these planets were ever tens of thousands of kilometers away, as he seems to think, the tides, both of water and of the solid body of our planet, would be hundreds of miles high. This is easily calculated from the height of the present water and body lunar tide, since the tide height is proportional to the mass of the tide-producing object and inversely proportional to the cube of the distance. To the best of my knowledge, there is no geological evidence for a global inundation of all parts of the world at any time between the sixth and fifteenth centuries B.C. If such floods had occurred, even if they were brief, they should have left some clear trace in the geological record. And what of the archaeological and paleontological evidence? Where are the extensive faunal extinctions of the correct date as a result of such floods? And where is the evidence of extensive melting in these centuries, near where the tidal distortion is greatest?

PROBLEM V
CHEMISTRY AND BIOLOGY OF THE TERRESTRIAL PLANETS

VELIKOVSKY's thesis has some peculiar biological and chemical consequences, which are compounded by some straightforward confusions on simple matters. He seems not to know (page 16) that oxygen is produced by green-plant photosynthesis on Earth. He makes no note of the fact that Jupiter is composed primarily of hydrogen and helium, while the atmosphere of Venus, which he supposes to have arisen inside of Jupiter, is composed almost entirely of carbon dioxide. These matters are central to his ideas and pose them very grave difficulties. Velikovsky holds that the manna that fell from the skies in the Sinai peninsula was of cometary origin and therefore that there are carbohydrates on both Jupiter and Venus. On the other hand, he quotes copious sources for fire and naphtha falling from the skies, which he interprets as celestial petroleum ignited in the Earth's oxidizing atmosphere (pages 53 through 58). Because Velikovsky believes in the reality and identity of both sets of events, his book displays a sustained confusion of carbohydrates and hydrocarbons; and at some points he seems to imagine that the Israelites were eating motor oil rather than divine nutriment during their forty years' wandering in the desert.

Reading the text is made still more difficult by the

apparent conclusion (page 366) of Martian polar caps made of manna, which are described ambiguously as "probably in the nature of carbon." Carbohydrates have a strong 3.5 micron infrared absorption feature, due to the stretching vibration of the carbon-hydrogen bond. No trace of this feature was observed in infrared spectra of the Martian polar caps taken by the Mariner 6 and 7 spacecraft in 1969. On the other hand, Mariner 6, 7 and 9 and Viking 1 and 2 have acquired abundant and persuasive evidence for frozen water and frozen carbon dioxide as the constituents of the polar caps.

Velikovsky's insistence on a celestial origin of petroleum is difficult to understand. Some of his references, for example in Herodotus, provide perfectly natural descriptions of the combustion of petroleum upon seepage to the surface in Mesopotamia and Iran. As Velikovsky himself points out (pages 55–56), the fire-rain and naphtha stories derive from precisely those regions of the Earth that have natural petroleum deposits. There is, therefore, a straightforward terrestrial explanation of the stories in question. The amount of downward seepage of petroleum in 2,700 years would not be very great. The difficulty in extracting petroleum from the Earth, which is the cause of certain practical problems today, would be greatly ameliorated if Velikovsky's hypothesis were true. It is also very difficult to understand on his hypothesis how it is, if oil fell from the skies in 1500 B.C., that petroleum deposits are intimately mixed with chemical and biological fossils of tens to hundreds of millions of years ago. But this circumstance is readily explicable if, as most geologists have concluded, petroleum arises from decaying vegetation, of the Carboniferous and other early geological epochs, and not from comets.

Even stranger are Velikovsky's views on extraterrestrial life. He believes that much of the "vermin," and particularly the flies referred to in *Exodus,* really fell from his comet—although he hedges on the extraterrestrial origin of frogs while approvingly quoting from the Iranian text, the *Bundahis* (page 183), which seems to admit a rain of cosmic frogs. Let us consider flies only.

Shall we expect houseflies or *Drosophila melanogaster* in forthcoming explorations of the clouds of Venus and Jupiter? He is quite explicit: "Venus—and therefore also Jupiter—is populated by vermin" (page 369). Will Velikovsky's hypothesis fall if no flies are found?

The idea that, of all the organisms on Earth, flies alone are of extraterrestrial origin is curiously reminiscent of Martin Luther's exasperated conclusion that, while the rest of life was created by God, the fly must have been created by the Devil because there is no conceivable practical use for it. But flies are perfectly respectable insects, closely related in anatomy, physiology and biochemistry to the other *insecta.* The possibility that 4.6 billion years of independent evolution on Jupiter—even if it were physically identical to the Earth—would produce a creature indistinguishable from other terrestrial organisms is to misread seriously the evolutionary process. Flies have the same enzymes, the same nucleic acids and even the same genetic code (which translates nucleic acid information into protein information) as do all the other organisms on Earth. There are too many intimate associations and identities between flies and other terrestrial organisms for them to have separate origins, as any serious investigation clearly shows.

In *Exodus,* Chapter 9, it is said that the cattle of Egypt all died, but of the cattle of the Children of Israel there "died not one." In the same chapter we find a plague that affects flax and barley but not wheat and rye. This fine-tuned host-parasite specificity is very strange for cometary vermin with no prior biological contact with Earth, but is readily explicable in terms of home-grown terrestrial vermin.

Then there is the curious fact that flies metabolize molecular oxygen. There is no molecular oxygen on Jupiter, nor can there be, because oxygen is thermodynamically unstable in an excess of hydrogen. Are we to imagine that the entire terminal electron transfer apparatus required for life to deal with molecular oxygen was adventitiously evolved on Jupiter by Jovian organisms hoping someday to be transported to Earth? This would be yet a bigger miracle than Velikovsky's principal collisional thesis. Velikovsky makes (page 187) a lame aside on the "ability of many small insects . . . to live in an atmosphere devoid of oxygen," which misses the point. The question is how an organism evolved on Jupiter could live in and metabolize an atmosphere rich in oxygen.

Next there is the problem of fly ablation. Small flies have just the same mass and dimensions as small meteors, which are burned up at an altitude of about 100 kilometers when they enter the Earth's atmosphere on cometary trajectories. Ablation accounts for the visibility of such meteors. Not only would cometary vermin be transformed rapidly into fried flies on entrance into the Earth's atmosphere; they would, as cometary meteors are today, be vaporized into atoms and never "swarm" over Egypt to the consternation of the Pharaoh. Likewise, the temperatures attendant to ejection of the comet from Jupiter, referred to above, would fry Velikovsky's flies. Impossible to begin with, doubly fried and atomized, cometary flies do not well survive critical scrutiny.

Finally, there is a curious reference to intelligent extraterrestrial life in *Worlds in Collision*. On page 364 Velikovsky argues that the near-collisions of Mars with Earth and Venus "make it highly improbable that any higher forms of life, if they previously existed there, survived on Mars." But when we examine the Mars as seen by Mariner 9 and Viking 1 and 2, we find that something over one-third of the planet has a modified cratered terrain somewhat reminiscent of the Moon and displays no sign of spectacular catastrophes other than ancient impacts. The other half to two-thirds of the planet shows almost no sign whatever of such impacts, but instead displays dramatic evidence of major tectonic activity, lava flows and vulcanism about a billion years ago. The small but detectable amount of impact cratering in this terrain shows that it was made much longer than several thousand years ago. There is no way to reconcile this picture with a view of a planet recently so devastated by impact catastrophism that all intelligent life would thereby have been eliminated. It is also by no means clear why, if all life on Mars were to be exterminated in such encounters, all life on Earth was not similarly exterminated.

PROBLEM VI

MANNA

MANNA, according to the etymology in *Exodus,* derives from the Hebrew words *man-hu,* which means "What is it?" Indeed, an excellent question! The idea of food falling from comets is not absolutely straightforward. Optical spectroscopy of comet tails, even before *Worlds in Collision* was published (1950), showed the presence of simple fragments of hydrocarbons, but no aldehydes—the building blocks of carbohydrates—were known then. They may nevertheless be present in comets. However, from the passage of Comet Kohoutek near the Earth, it is now known that comets contain large quantities of simple nitriles—in particular, hydrogen cyanide and methyl cyanide. These are poisons, and it is not immediately obvious that comets are good to eat.

But let us put this objection aside, grant Velikovsky his hypothesis, and calculate the consequences. How much manna is required to feed the hundreds of thousands of Children of Israel for forty years (see *Exodus,* Chapter 16, Verse 35)?

In *Exodus,* Chapter 16, Verse 20, we find that the manna left overnight was infested by worms in the morning—an event possible with carbohydrates but extremely unlikely with hydrocarbons. Moses may have been a better chemist than Velikovsky. This event also shows that manna was not storable. It fell every day for forty years according to the Biblical account. We will assume that the quantity that fell every day was just sufficient to feed the Children of Israel, although Velikovsky assures us (page 138) from Midrashic sources that the quantity that fell was adequate for two thousand years rather than a mere forty. Let us assume that each Israelite ate on the order of a third of a kilogram of manna per day, somewhat less than a subsistence diet. Then each will eat 100 kilograms per year and 4,000 kilograms in forty years. Hundreds of thousands of Israelites, the number explicitly mentioned in *Exodus,* will then consume something over a million kilograms of manna during the forty years' wandering in the desert. But we cannot imagine the debris from the cometary tail falling each day,* preferentially on the portion of the Wilderness of Sin in which the Israelites happened to have wandered. This would be no less miraculous than the Biblical account taken at face value. The area occupied by a few hundred thousand itinerant tribesmen, wandering under a common leadership, is, very roughly, several times 10^{-7} the area of the Earth. Therefore, during the forty years of wandering, all of the Earth must have accumulated several times 10^{18} grams of manna, or enough to cover the entire surface of the planet with manna to a depth of about an inch. If this indeed happened, it would certainly be a memorable event, and may even account for the gingerbread house in "Hansel and Gretel."

Now, there is no reason for the manna to have fallen only on Earth. In forty years the tail of the comet, if restricted to the inner solar system, would have traversed some 10^{10} km. Making only a modest allowance for the ratio of the volume of the Earth to the volume of the tail, we find that the mass of manna distributed to the inner solar system by this event is larger than 10^{28} grams. This is not only more massive by many orders of magnitude than the most massive comet known; it is

---

* Actually, *Exodus* states that manna fell each day except on the Sabbath. A double ration, uninfected by worms, fell instead on Friday. This seems awkward for Velikovsky's hypothesis. How could the comet know? Indeed, this raises a general problem about Velikovsky's historical method. Some quotations from his religious and historical sources are to be taken literally; others are to be dismissed as "local embellishments." But what is the standard by which this decision is made? Surely such a standard must involve a criterion independent of our predispositions toward Velikovsky's contentions.

already more massive than the planet Venus. But comets cannot be composed only of manna. (Indeed, no manna at all has been detected so far in comets.) Comets are known to be composed primarily of ices, and a conservative estimate of the ratio of the mass of the comet to the mass of the manna is much larger than 10^3. Therefore, the mass of the comet must be much larger than 10^{31} grams. This is the mass of Jupiter. If we were to accept Velikovsky's Midrashic source above, we would deduce that the comet had a mass comparable to that of the Sun. Interplanetary space in the inner solar system should even today be filled with manna. I leave it to the reader to make his own judgment on the validity of Velikovsky's hypothesis in the light of such calculations.

PROBLEM VII

THE CLOUDS OF VENUS

VELIKOVSKY's prognostication that the clouds of Venus were made of hydrocarbons or carbohydrates has many times been hailed as an example of a successful scientific prediction. From Velikovsky's general thesis and the calculations just described above, it is clear that Venus should be saturated with manna, a carbohydrate. Velikovsky says (page x) that "the presence of hydrocarbon gases and dust in the cloud envelope of Venus would constitute a crucial test" for his ideas. It is also not clear whether "dust" in the foregoing quotation refers to hydrocarbon dust or just ordinary silicate dust. On the same page Velikovsky quotes himself as saying, "On the basis of this research, I assume that Venus must be rich in petroleum gases," which seems to be an unambiguous reference to the components of natural gas, such as methane, ethane, ethylene and acetylene.

At this point, a little history must enter our story. In the 1930s and early 1940s, the only astronomer in the world concerning himself with planetary chemistry was the late Rupert Wildt, once of Göttingen, and later at Yale. It was Wildt who first identified methane in the atmospheres of Jupiter and Saturn, and it was he who first proposed the presence of higher hydrocarbon gases in the atmospheres of these planets. Thus, the idea that "petroleum gases" might exist on Jupiter is not original with Velikovsky. Likewise, it was Wildt who proposed that formaldehyde might be a constituent of the atmosphere of Venus, and that a carbohydrate polymer of formaldehyde might constitute the clouds. The idea of carbohydrates in the clouds of Venus was not original with Velikovsky either, and it is difficult to believe that one who so thoroughly researched the astronomical literature of the 1930s and 1940s was unaware of these papers by Wildt which relate so closely to Velikovsky's central theme. Yet there is no mention whatever of the Jupiter phase of Wildt's work and only a footnote on formaldehyde (page 368), without references, and without any acknowledgment that Wildt had proposed carbohydrates on Venus. Wildt, unlike Velikovsky, understood well the difference between hydrocarbons and carbohydrates; moreover, he performed unsuccessful spectroscopic searches in the near-ultraviolet for the proposed formaldehyde monomer. Being unable to find the monomer, he abandoned the hypothesis in 1942. Velikovsky did not.

As I pointed out many years ago (Sagan, 1961), the vapor pressure of simple hydrocarbons in the vicinity of the clouds of Venus should make them detectable if they comprise the clouds. They were not detectable then, and in the intervening years, despite a wide range of analytic techniques used, neither hydrocarbons nor carbohydrates have been found. These molecules have been searched for by high-resolution ground-based optical spectroscopy, including Fourier transform techniques; by ultraviolet spectroscopy from the Wisconsin Experimental Package of the Orbiting Astronomical Observatory OAO-2; by ground-based infrared observations; and by direct entry probes of the Soviet Union and the United States. Not one of them has been found. Typical abundance upper limits on the simplest hydrocarbons and on aldehydes, the building blocks of carbohydrates, are a few parts per million (Connes, *et al.*,

1967; Owen and Sagan, 1972). [The corresponding upper limits for Mars are also a few parts per million (Owen and Sagan, 1972)]. All observations are consistent in showing that the bulk of the Venus atmosphere is composed of carbon dioxide. Indeed, because the carbon is present in such an oxidized form, at best trace constituents of the simple reduced hydrocarbons could be expected. Observations on the wings of the critical 3.5 micron region show not the slightest trace of the C-H absorption feature common to both hydrocarbons and carbohydrates (Pollack, *et al.*, 1974). All other absorption bands in the Venus spectrum, from the ultraviolet through the infrared, are now understood; none of them is due to hydrocarbons or carbohydrates. No specific organic molecule has ever been suggested that can explain with precision the infrared spectrum of Venus as it is now known.

Moreover, the question of the composition of the Venus clouds—a major enigma for centuries—was solved not long ago (Young and Young, 1973; Sill, 1972; Young, 1973; Pollack, *et al.*, 1974). The clouds of Venus are composed of an approximately 75 percent solution of sulfuric acid. This identification is consistent with the chemistry of the Venus atmosphere, in which hydrofluoric and hydrochloric acid have also been found; with the real part of the refractive index, deduced from polarimetry, which is known to three significant figures (1.44); with the 11.2 micron and 3 micron (and, now, far-infrared) absorption features; and with the discontinuity in the abundance of water vapor above and below the clouds. These observed features are inconsistent with the hypothesis of hydrocarbon or carbohydrate clouds.

With such organic clouds now so thoroughly discredited, why do we hear about space-vehicle research having corroborated Velikovsky's thesis? This also requires a story. On December 14, 1962, the first successful American interplanetary spacecraft, Mariner 2, flew by Venus. Built by the Jet Propulsion Laboratory, it carried, among other more important instruments, an infrared radiometer for which I happened to be one of four experimenters. This was at a time before even the first successful lunar Ranger spacecraft, and NASA was comparatively inexperienced in releasing the scientific findings. A press conference was held in Washington to announce the results, and Dr. L. D. Kaplan, one of the experimenters on our team, was delegated to describe the results to the assembled reporters. It is clear that when his time came, he described the results with somewhat the following flavor (these are not his exact words): "Our experiment was a two-channel infrared radiometer, one channel centered in the 10.4 micron CO_2 hot band, the other in an 8.4 micron clear window in the gas phase of the Venus atmosphere. The objective was to measure absolute brightness temperatures and differential transmission between the two channels. A limb-darkening law was found in which the normalized intensity varied as mu to the power alpha, where mu is the arccosine of the angle between the local planetary normal and the line of sight, and—"

At some such point he was interrupted by impatient reporters, unused to the intricacies of science, who said something like "Don't tell us the dull stuff; give us the real poop! How thick are the clouds, how high are they, and what are they made of?" Kaplan replied, quite properly, that the infrared radiometer experiment was not designed to test such questions, nor did it. But then he said something like "I'll tell you what I think." He went on to describe his view that the greenhouse effect, in which an atmosphere is transparent to visible sunlight but opaque to infrared emission from the surface, needed to keep the surface of Venus hot, might not work on Venus because the atmospheric constituents seemed to be transparent at a wavelength in the vicinity of 3.5 microns. If some absorber at this wavelength existed in the Venus atmosphere, the window could be plugged, the greenhouse effect retained, and the high surface temperature accounted for. He proposed that hydrocarbons would be splendid greenhouse molecules.

Kaplan's cautions were not noted by the press, and the next day headlines could be found in many American newspapers saying: "Hydrocarbon Clouds Found on Venus by Mariner 2." Meanwhile, back at the Jet Propulsion Laboratory, several Laboratory publicists were in the process of writing a popular report on the mission, since called "Mariner: Mission to Venus." One imagines them in the midst of writing, picking up the morning newspaper and saying, "Hey! I didn't know we found hydrocarbon clouds on Venus." And, indeed, that publication lists hydrocarbon clouds as one of the principal discoveries of Mariner 2: "At their base, the clouds are about 200 degrees F and probably are comprised of condensed hydrocarbons held in oily suspension." (The report also opts for greenhouse heating of the Venus surface, but Velikovsky has chosen to believe only a part of what was printed.)

One now imagines the Administrator of NASA passing on the good tidings to the President in the annual report of the Space Administration; the President handing it on yet another step in his annual report to Congress; and the writers of elementary astronomy texts, always anxious to include the very latest results, enshrining this "finding" in their pages. With so many apparently reliable, high-level and mutually consistent reports that Mariner 2 found hydrocarbon clouds on Venus, it is no wonder that Velikovsky and several fair-minded scientists, inexperienced in the mysterious ways of NASA, might deduce that here is the classic test of a scientific theory: an apparently bizarre prediction, made before the observation, and then unexpectedly confirmed by experiment.

The true situation is very different, as we have seen. Neither Mariner 2 nor any subsequent investigation of the Venus atmosphere has found evidence for hydrocarbons or carbohydrates, in gas, liquid or solid phase. It is now known (Pollack, 1969) that carbon dioxide and water vapor adequately fill the 3.5 micron window. The Pioneer Venus mission in late 1978 found just the water vapor needed, along with the long-observed quantity of carbon dioxide, to account for the high surface temperature through the greenhouse effect. It is ironic that the Mariner 2 "argument" for hydrocarbon clouds on Venus in fact derives from an attempt to rescue the greenhouse explanation of the high surface temperature, which Velikovsky does not support. It is also ironic that Professor Kaplan was later a co-author of a paper that established a very low abundance of methane, a "petroleum gas," in a spectroscopic examination of the Venus atmosphere (Connes, *et al.*, 1967).

In summary, Velikovsky's idea that the clouds of Venus are composed of hydrocarbons or carbohydrates is neither original nor correct. The "crucial test" fails.

PROBLEM VIII

THE TEMPERATURE OF VENUS

ANOTHER CURIOUS circumstance concerns the surface temperature of Venus. While the high temperature of Venus is often quoted as a successful prediction and a support of Velikovsky's hypothesis, the reasoning behind his conclusion and the consequences of his arguments do not seem to be widely known nor discussed.

Let us begin by considering Velikovsky's views on the temperature of Mars (pages 367–368). He believes that Mars, being a relatively small planet, was more severely affected in its encounters with the more massive Venus and Earth, and therefore that Mars should have a high temperature. He proposes that the mechanism may be "a conversion of motion into heat," which is a little vague, since heat is precisely the motion of molecules or, much more fantastic, by "interplanetary electrical discharges" which "could also initiate atomic fissions with ensuing radioactivity and emission of heat."

In the same section, he baldly states, "Mars emits more heat than it receives from the Sun," in apparent consistency with his collision hypothesis. This statement is, however, dead wrong. The temperature of Mars has been measured repeatedly by Soviet and American spacecraft and by ground-based observers, and the temperatures of all parts of Mars are just what is calculated from the amount of sunlight absorbed by the surface. What is more, this was well known in the 1940s, before Velikovsky's book was published. And while he mentions four prominent scientists who were involved before 1950 in measuring the temperature of Mars, he makes no reference to their work, and explicitly and erroneously states that they concluded that Mars gave off more radiation than it received from the Sun.

It is difficult to understand this set of errors, and the most generous hypothesis I can offer is that Velikovsky confused the visible part of the electromagnetic spectrum, in which sunlight heats Mars, with the infrared part of the spectrum, in which Mars largely radiates to space. But the conclusion is clear. Mars, even more than Venus, by Velikovsky's argument should be a "hot planet." Had Mars proved to be unexpectedly hot, perhaps we would have heard of this as a further confirmation of Velikovsky's views. But when Mars turns out to have exactly the temperature everyone expected it to have, we do not hear of this as a refutation of Velikovsky's views. There is a planetary double standard at work.

When we now move on to Venus, we find rather similar arguments brought into play. I find it odd that Velikovsky does not attribute the temperature of Venus to its ejection from Jupiter (see Problem I, above), but he does not. Instead, we are told, because of its close encounter with the Earth and Mars, Venus must have been heated, but also (page 77) "the head of the comet . . . had passed close to the Sun and was in a state of candescence." Then, when the comet became the planet Venus, it must still have been "very hot" and have "given off heat" (page ix). Again pre-1950 astronomical observations are referred to (page 370), which show that the dark side of Venus is approximately as hot as the bright side of Venus, to the level probed by middle-infrared radiation. Here Velikovsky accurately quotes the astronomical investigators, and from their work deduces (page 371) "the night side of Venus radiates heat because Venus is hot." Of course!

What I think Velikovsky is trying to say here is that

his Venus, like his Mars, is giving off more heat than it receives from the Sun, and that the observed temperatures on both the night and day sides are due more to the "candescence" of Venus than to the radiation it now receives from the Sun. But this is a serious error. The bolometric albedo (the fraction of sunlight reflected by an object at all wavelengths) of Venus is about 0.73, entirely consistent with the observed infrared temperature of the clouds of Venus of about 240°K; that is to say, the clouds of Venus are precisely at the temperature expected on the basis of the amount of sunlight that is absorbed there.

Velikovsky proposed that both Venus and Mars give off more heat than they receive from the Sun. He is wrong for both planets. In 1949 Kuiper (see References) suggested that Jupiter gives off more heat than it receives, and subsequent observations have proved him right. But of Kuiper's suggestion *Worlds in Collision* breathes not a word.

Velikovsky proposed that Venus is hot because of its encounters with Mars and the Earth, and its close passage to the Sun. Since Mars is not anomalously hot, the high surface temperature of Venus must be attributed primarily to the passage of Venus near the Sun during its cometary incarnation. But it is easy to calculate how much energy Venus would have received during its close passage to the Sun and how long it would take for this energy to be radiated away into space. This calculation is performed in Appendix 3, where we find that all of this energy is lost in a period of months to years after the close passage to the Sun, and that there is no chance of any of that heat being retained at the present time in Velikovsky's chronology. Velikovsky does not mention how close to the Sun Venus is supposed to have passed, but a very close passage compounds the already extremely grave collision physics difficulties outlined in Appendix 1. Incidentally, there is a slight hint in *Worlds in Collision* that Velikovsky believes that comets shine by emitted rather than reflected light. If so, this may be the source of some of his confusion regarding Venus.

Velikovsky nowhere states the temperature he believed Venus to be at in 1950. As mentioned above, on page 77 he says vaguely that the comet that later became Venus was in a state of "candescence," but in the preface to the 1965 edition (page xi), he claims to have predicted "an incandescent state of Venus." This is not at all the same thing, because of the rapid cooling after its supposed solar encounter (Appendix 3). Moreover, Velikovsky himself is proposing that Venus is cooling through time, so what precisely Velikovsky meant by saying that Venus is "hot" is to some degree obscure.

Velikovsky writes in the 1965 preface that his claim of a high surface temperature was "in total disagreement with what was known in 1946." This turns out to be not quite the case. The dominant figure of Rupert Wildt again looms over the astronomical side of Velikovsky's hypothesis. Wildt, who, unlike Velikovsky, understood the nature of the problem, predicted correctly that Venus and not Mars would be "hot." In a 1940 paper in the *Astrophysical Journal,* Wildt argued that the surface of Venus was much hotter than conventional astronomical opinion had held, because of a carbon-dioxide greenhouse effect. Carbon dioxide had recently been discovered spectroscopically in the atmosphere of Venus, and Wildt correctly pointed out that the observed large quantity of CO_2 would trap infrared radiation given off by the surface of the planet until the surface temperature rose to a higher value, so that the incoming visible sunlight just balanced the outgoing infrared planetary emission. Wildt calculated that the temperature would be almost 400°K, or around the normal boiling point of water (373°K = 212°F = 100°C). There is no doubt that this was the most careful treatment of the surface temperature of Venus prior to the 1950s, and it is again odd that Velikovsky, who seems to have read all papers on Venus and Mars published in the *Astrophysical Journal* in the 1920s, 1930s and 1940s, somehow overlooked this historically significant work.

We now know from ground-based radio observations

and from the remarkably successful direct entry and landing probes of the Soviet Union that the surface temperature of Venus is within a few degrees of 750°K (Marov, 1972). The surface atmospheric pressure is about ninety times that at the surface of the Earth, and is comprised primarily of carbon dioxide. This large abundance of carbon dioxide, plus the smaller quantities of water vapor which have been detected on Venus, are adequate to heat the surface to the observed temperature via the greenhouse effect. The Venera 8 descent module, the first spacecraft to land on the illuminated hemisphere of Venus, found it illuminated at the surface, and the Soviet experimenters concluded that the amount of sunlight reaching the surface and the atmospheric constitution were together adequate to drive the required radiative-convective greenhouse (Marov, *et al.,* 1973). These results were confirmed by the Venera 9 and 10 missions, which obtained clear photographs, in sunlight, of surface rocks. Velikovsky is thus certainly mistaken when he says (page ix) "light does not penetrate the cloud cover," and is probably mistaken when he says (page ix) the "greenhouse effect could not explain so high a temperature." These conclusions received important additional support late in 1978 from the U.S. Pioneer Venus mission.

A repeated claim by Velikovsky is that Venus is cooling off with time. As we have seen, he attributes its high temperature to solar heating during a close solar passage. In many publications Velikovsky compares published temperature measurements of Venus, made at different times, and tries to show the desired cooling. An unbiased presentation of the microwave brightness temperatures of Venus—the only nonspacecraft data that apply to the surface temperature of the planet—are exhibited in Figure 1. The error bars represent the uncertainties in the measurement processes as estimated by the radio observers themselves. We see that there is not the faintest hint of a decline in temperature with time (if anything, there is a suggestion of an increase with time, but the error bars are sufficiently large that such a conclusion is also unsupported by the data).

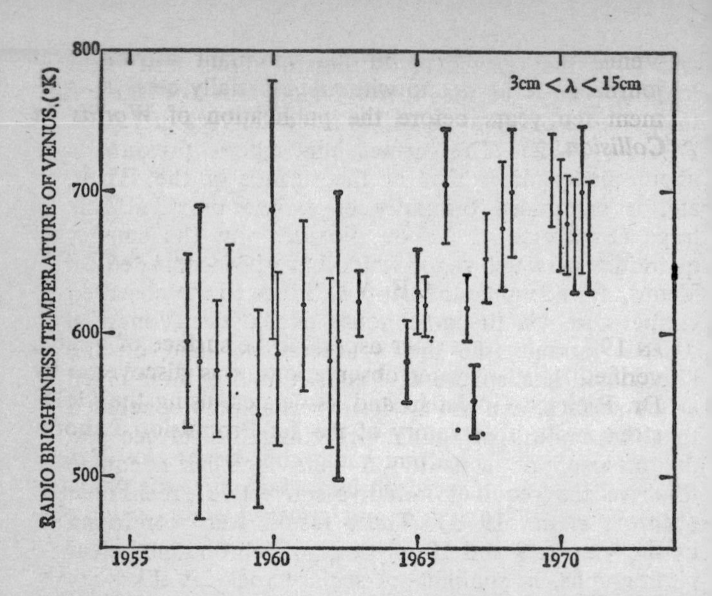

FIGURE 1. *Microwave brightness temperatures of Venus as a function of time (after a compilation by D. Morrison). There is certainly no evidence of a declining surface temperature. The wavelength of observation is denoted by λ.*

Similar results apply to measurements, in the infrared part of the spectrum, of cloud temperatures: they are lower in magnitude and do not decline with time. Moreover, the simplest considerations of the solution of the one-dimensional equation of heat conduction show that in the Velikovskian scenario essentially all the cooling by radiation to space would have occurred long ago. Even if Velikovsky were right about the source of the high Venus surface temperatures, his prediction of a secular temperature decrease would be erroneous.

The high surface temperature of Venus is another of the so-called proofs of the Velikovsky hypothesis. We find that (1) the temperature in question was never specified; (2) the mechanism proposed for providing this temperature is grossly inadequate; (3) the surface of the planet does not cool off with time as advertised; and (4) the idea of a high surface temperature on

Venus was published in the dominant astronomical journal of its time and with an essentially correct argument ten years before the publication of *Worlds in Collision.*

PROBLEM IX

THE CRATERS AND MOUNTAINS OF VENUS

IN 1973 AN IMPORTANT aspect of the surface of Venus, verified by many later observations, was discovered by Dr. Richard Goldstein and associates, using the Goldstone radar observatory of the Jet Propulsion Laboratory. They found, from radar that penetrates Venus' clouds and is reflected off its surface, that the planet is mountainous in places and cratered abundantly; perhaps, like parts of the Moon, saturation-cratered—i.e., so packed with craters that one crater overlaps the other. Because successive volcanic eruptions tend to use the same lava tube, saturation cratering is more characteristic of impact than of volcanic cratering mechanisms. This is not a conclusion predicted by Velikovsky, but that is not my point. These craters, like the craters in the lunar maria (plural for Latin *mare,* "sea"), on Mercury and in the cratered regions of Mars, are produced almost exclusively by the impact of interplanetary debris. Large crater-forming objects are not dissipated as they enter the Venus atmosphere, despite its high density. Now, the colliding objects cannot have arrived at Venus in the past ten thousand years; otherwise, the Earth would be as plentifully cratered. The most likely source of these collisions is the Apollo objects (asteroids whose orbits cross the orbit of the Earth) and small comets we have already discussed (Appendix 1). But for them to produce as many craters as Venus possesses, the cratering process on Venus must have taken billions of years. Alternatively, the cratering may have occurred more rapidly in the very earliest history of the solar system, when interplanetary debris was much more plentiful. But there is no way for it to have happened recently. On the other hand, if Venus was, several thousand years ago, in the deep interior of Jupiter, there is no way it could have accumulated such impacts there. The clear conclusion from the craters of Venus is, therefore, that Venus has for billions of years been an object exposed to interplanetary collisions—in direct contradiction to the fundamental premise of Velikovsky's hypothesis.

The Venus craters are significantly eroded. Some of the rocks on the surface of the planet, as revealed by the Venera 9 and 10 photography, are quite young; others are severely eroded. I have described elsewhere possible mechanisms for erosion on the Venus surface—including chemical weathering and slow deformation at high temperatures (Sagan, 1976). However, these findings have no bearing whatever on the Velikovskian hypotheses: recent volcanic activity on Venus need no more be attributed to a close passage to the Sun or to Venus' being in some vague sense a "young" planet than recent volcanic activity on Earth.

In 1967 Velikovsky wrote: "Obviously, if the planet is billions of years old, it could not have preserved its original heat; also, any radioactive process that can produce such heat must be of a very rapid decay [*sic*], and this again would not square with an age of the planet counted in billions of years." Unfortunately, Velikovsky has failed to understand two classic and basic geophysical results. Thermal conduction is a much slower process than radiation or convection, and, in the case of the Earth, primordial heat makes a detectable contribution to the geothermal temperature gradient and to the heat flux from the Earth's interior. The same applies to Venus. Also, the radionuclides responsible for radioactive heating of the Earth's crust are long-lived isotopes of uranium, thorium and potassium—isotopes with half-lives comparable to the age of the planet. Again, the same applies to Venus.

If, as Velikovsky believes, Venus were completely molten only a few thousand years ago—from planetary collisions or any other cause—no more than a thin outer crust, at most $\sim$ 100 meters thick, could since have been produced by conductive cooling. But the radar observations reveal enormous linear mountain ranges, ringed basins, and a great rift valley, with dimensions of hundreds to thousands of kilometers. It is very unlikely that such extensive tectonic or impact features could be stably supported over a liquid interior by such a thin and fragile crust.

PROBLEM X

THE CIRCULARIZATION OF THE ORBIT OF VENUS AND NONGRAVITATIONAL FORCES IN THE SOLAR SYSTEM

THE IDEA that Venus could have been converted, in a few thousand years, from an object in a highly elongated or eccentric orbit to its present orbit, which is—except for Neptune—the most nearly perfect circular orbit of all the planets, is at odds with what we know about the three-body problem* in celestial mechanics. However, it must be admitted that this is not a completely solved problem, and that, while the odds are large, they are not absolutely overwhelming against Velikovsky's hypothesis on this score. Furthermore, when Velikovsky invokes electrical or magnetic forces, with no effort to calculate their magnitude or describe in detail their effects, we are hard pressed to assess his ideas. However, simple arguments from the required magnetic energy density to circularize a comet show that the field strengths implied are unreasonably high (Appendix 4)—they are counterindicated by studies of rock magnetization.

We can also approach the problem empirically. Straightforward Newtonian mechanics is able to predict with remarkable accuracy the trajectories of spacecraft—so that, for example, the Viking orbiters were placed within 100 kilometers of their designated orbit;

* The prediction of the relative motions of three objects attracted to each other gravitationally.

Venera 8 was placed precisely on the sunlit side of the equatorial terminator of Venus; and Voyager 1 was placed in exactly the correct entry corridor in the vicinity of Jupiter to be directed close to Saturn. No mysterious electrical or magnetic influences were encountered. Newtonian mechanics is adequate to predict, with great precision, for example, the exact moment when the Galilean satellites of Jupiter will eclipse each other.

Comets, it is true, have somewhat less predictable orbits, but this is almost certainly because there is a boiling off of frozen ices as these objects approach the Sun, and a small rocket effect. The cometary incarnation of Venus, if it existed, might also have had such icy vaporization, but there is no way in which the rocket effect would have preferentially brought that comet into close passages with the Earth or Mars. Halley's comet, which has probably been observed for two thousand years, remains on a highly eccentric orbit and has not been observed to show the slightest tendency toward circularization; yet it is almost as old as Velikovsky's "comet." It is extraordinarily unlikely that Velikovsky's comet, had it ever existed, became the planet Venus.

SOME OTHER PROBLEMS

THE PRECEDING ten points are the major scientific flaws in Velikovsky's argument, as nearly as I can determine. I have discussed earlier some of the difficulties with his approach to ancient writings. Let me here list a few of the miscellaneous other problems I have encountered in reading *Worlds in Collision*.

On page 280 the Martian moons Phobos and Deimos are imagined to have "snatched some of Mars' atmosphere" and to thereby appear very bright. But it is immediately clear that the escape velocity on these objects—perhaps 20 miles per hour—is so small as to make them incapable of retaining even temporarily any atmosphere; close-up Viking photographs show no atmosphere and no frost patches; and they are among the darkest objects in the solar system.

Beginning on page 281, there is a comparison of the Biblical Book of Joel and a set of Vedic hymns describing "maruts." Velikovsky believes that the "maruts" were a host of meteorites that preceded and followed Mars during its close approach to Earth, which he also believes is described in Joel. Velikovsky says (page 286): "Joel did not copy from the Vedas nor the Vedas from Joel." Yet, on page 288, Velikovsky finds it "gratifying" to discover that the words "Mars" and "marut" are cognates. But how, if the stories in Joel and the Vedas are independent, could the two words possibly be cognates?

On page 307 we find Isaiah making an accurate prediction of the time of the return of Mars for another collision with Earth "based on experience during previous perturbations." If so, Isaiah must have been able to solve the full three-body problem with electrical and magnetic forces thrown in, and it is a pity that this knowledge was not also passed down to us in the Old Testament.

On pages 366 and 367 we find an argument that Venus, Mars and Earth, in their interactions, must have exchanged atmospheres. If massive quantities of terrestrial molecular oxygen (20 percent of our atmosphere) were transferred to Mars and Venus 3,500 years ago, they should be there still in massive amounts. The time scale for turnover of O_2 in the Earth's atmosphere is 2,000 years, and that is by a biological process. In the absence of abundant biological respiration, any O_2 on Mars and Venus 3,500 years ago should still be there. Yet we know quite definitely from spectroscopy that O_2 is at best a tiny constituent of the already extremely thin Martian atmosphere (and is likewise scarce on Venus). Mariner 10 found evidence of oxygen in the atmosphere of Venus—but tiny quantities of atomic oxygen in the upper atmosphere, not massive quantities of molecular oxygen in the lower atmosphere.

The dearth of O_2 on Venus also renders untenable

Velikovsky's belief in petroleum fires in the lower Venus atmosphere—neither the fuel nor the oxidant is present in appreciable amounts. These fires, Velikovsky believed, would produce water, which would be photodissociated, yielding O. Thus Velikovsky requires significant deep atmospheric O_2 to account for upper atmospheric O. In fact, the O found is understood very well in terms of the photochemical breakdown of the principal atmospheric constituent, CO_2, into CO and O. These distinctions seem to have been lost on some of Velikovsky's supporters, who seized on the Mariner 10 findings as a vindication of *Worlds in Collision.*

Since there is negligible oxygen and water vapor in the Martian atmosphere, Velikovsky argues, some other constituent of the Martian atmosphere must be derived from the Earth. The argument, unfortunately, is a *non sequitur*. Velikovsky opts for argon and neon, despite the fact that these are quite rare constituents of the Earth's atmosphere. The first published argument for argon and neon as major constituents of the Martian atmosphere was made by Harrison Brown in the 1940s. More than trace quantities of neon are now excluded; about one percent argon was found by Viking. But even if large quantities of argon had been found on Mars, it would have provided no evidence for a Velikovskian atmospheric exchange—because the most abundant form of argon, ^{40}Ar, is produced by the radioactive decay of potassium 40, which is expected in the crust of Mars.

A much more serious problem for Velikovsky is the relative absence of N_2 (molecular nitrogen) from the Martian atmosphere. The gas is relatively unreactive, does not freeze out at Martian temperatures and cannot rapidly escape from the Martian exosphere. It is the major constituent of the Earth's atmosphere but comprises only one percent of the Martian atmosphere. If such an exchange of gases occurred, where is all the N_2 on Mars? These tests of the assumed gas exchange between Mars and the Earth, which Velikovsky advocates, are poorly thought out in his writings; and the tests contradict his thesis.

Worlds in Collision is an attempt to validate Biblical and other folklore as history, if not theology. I have tried to approach the book with no prejudgments. I find the mythological concordances fascinating, and worth further investigation, but they are probably explicable on diffusionist or other grounds. The scientific part of the text, despite all the claims of "proofs," runs into at least ten very grave difficulties.

Of the ten tests of Velikovsky's work described above, there is not one case where his ideas are simultaneously original and consistent with simple physical theory and observation. Moreover, many of the objections—especially Problems I, II, III and X—are objections of high weight, based on the motion and conservation laws of physics. In science, an acceptable argument must have a clearly set forth chain of evidence. If a single link in the chain is broken, the argument fails. In the case of *Worlds in Collision,* we have the opposite case: virtually every link in the chain is broken. To rescue the hypothesis requires special pleading, the vague invention of new physics, and selective inattention to a plethora of conflicting evidence. Accordingly, Velikovsky's basic thesis seems to me clearly untenable on physical grounds.

Moreover, there is a dangerous potential problem with the mythological material. The supposed events are reconstructed from legends and folktales. But these global catastrophes are not present in the historical records or folklore of many cultures. Such strange omissions are accounted for, when they are noted at all, by "collective amnesia." Velikovsky wants it both ways. Where concordances exist, he is prepared to draw the most sweeping conclusions from them. Where concordances do not exist, the difficulty is dismissed by invoking "collective amnesia." With so lax a standard of evidence, anything can be "proved."

I should also point out that a much more plausible explanation exists for most of the events in *Exodus* that Velikovsky accepts, an explanation that is much more in accord with physics. The Exodus is dated in *I Kings* as occurring 480 years before the initiation of the construction of the Temple of Solomon. With other supporting calculations, the date for the Biblical Exodus is then computed to be about 1447 B.C. (Covey, 1975). Other Biblical scholars disagree, but this date is consistent with Velikovsky's chronology, and is astonishingly close to the dates obtained by a variety of scientific methods for the final and colossal volcanic explosion of the island of Thera (or Santorin) which may have destroyed the Minoan civilization in Crete and had profound consequences for Egypt, less than three hundred miles to the south. The best available radiocarbon date for the event, obtained from a tree buried in volcanic ash on Thera, is 1456 B.C. with an error in the method of at least plus or minus forty-three years. The amount of volcanic dust produced is more than adequate to account for three days of darkness in daytime, and accompanying events can explain earthquakes, famine, vermin and a range of familiar Velikovskian catastrophes. It also may have produced an immense Mediterranean *tsunami*, or tidal wave, which Angelos Galanopoulos (1964)—who is responsible for much of the recent geological and archaeological interest in Thera—believes can account for the parting of the Red Sea as well.* In a certain sense, the Galanopoulos explanation of the events in *Exodus* is even more provocative than the Velikovsky explanation, because Galanopoulos has presented moderately convincing evidence that Thera corresponds in almost all essential details to the legendary civilization of Atlantis. If he is right, it is the destruction of Atlantis rather than the apparition of a comet that permitted the Israelites to leave Egypt.

There are many strange inconsistencies in *Worlds in Collision*, but on the next-to-last page of the book,

* An informative and entertaining discussion of the Thera case, and the whole question of the connection of myth with geological events, can be found in the book by Vitaliano (1973); see also de Camp (1975).

a breathtaking departure from the fundamental thesis is casually introduced. We read of a hoary and erroneous analogy between the structures of solar systems and of atoms. Suddenly we are presented with the hypothesis that the supposed errant motions of the planets, rather than being caused by collisions, are instead the result of changes in the quantum energy levels of planets attendant to the absorption of a photon—or perhaps several. Solar systems are held together by gravitational forces; atoms by electrical forces. While both forces depend on the inverse square of distance, they have totally different characters and magnitudes: as one of many differences, there are positive and negative electrical charges, but only one sign of gravitational mass. We understand both solar systems and atoms well enough to see that Velikovsky's proposed "quantum jumps" of planets are based on a misunderstanding of both theories and evidence.

To the best of my knowledge, in *Worlds in Collision* there is not a single correct astronomical prediction made with sufficient precision for it to be more than a vague lucky guess—and there are, as I have tried to point out, a host of demonstrably false claims. The existence of strong radio emission from Jupiter is sometimes pointed to as the most striking example of a correct prediction by Velikovsky, but all objects give off radio waves if they are at temperatures above absolute zero. The essential characteristics of the Jovian radio emission—that it is nonthermal, polarized, intermittent radiation, connected with the vast belts of charged particles which surround Jupiter, trapped by its strong magnetic field—are nowhere predicted by Velikovsky. Further, his "prediction" is clearly not linked in its essentials to the fundamental Velikovskian theses.

Merely guessing something right does not necessarily demonstrate prior knowledge or a correct theory. For example, in an early science-fiction work dated 1949, Max Ehrlich imagined a near-collision of the Earth with another cosmic object, which filled the sky and terrorized the inhabitants of the Earth. Most frightening was the fact that on this passing planet was a natural feature which looked very much like a huge eye. This is one of many fictional and serious antecedents to Velikovsky's idea that such collisions happen frequently. But that is not my point. In a discussion of how it is that the side of the Moon facing the Earth has large smooth maria while the averted face of the Moon is almost free of them, John Wood of the Smithsonian Astrophysical Observatory proposed that the side of the Moon now turned toward the Earth was once at the edge, or limb, of the Moon, on the leading hemisphere of the Moon's motion about the Earth. In this position it swept up, billions of years ago, a ring of debris which surrounded the Earth and which may have been involved in the formation of the Earth-Moon system. By Euler's laws, the Moon must then have altered its rotation axis to correspond to its new principal moment of inertia, so that its leading hemisphere then faced the Earth. The remarkable conclusion is that there would have been a time, according to Wood, when what is now the eastern limb of the Moon would have been facing the Earth. But the eastern limb of the Moon has an enormous collision feature, billions of years old, called Mare Orientale, which looks very much like a giant eye. No one has suggested that Ehrlich was relying upon a racial memory of an event three billion years old when he wrote *The Big Eye*. It is merely a coincidence. When enough fiction is written and enough scientific hypotheses are proposed, sooner or later there will be accidental concordances.

With these enormous liabilities, how is it that *Worlds in Collision* has been so popular? Here I can only guess. For one thing, it is an attempted validation of religion. The old Biblical stories are literally true, Velikovsky tells us, if only we interpret them in the right way. The Jewish people, for example, saved from Egyptian Pharaohs, Assyrian kings and innumerable other disasters by obliging cometary intervention, had every right, he seems to be saying, to believe themselves chosen. Velikovsky attempts to rescue not only religion but also astrology: the outcomes of wars, the fates of whole peoples, are determined by the positions of the planets. In some sense, his work holds out a promise of the cosmic connectedness of mankind—a sentiment with which I sympathize, but in a somewhat different context *(The Cosmic Connection)*—and the reassurance that ancient peoples and other cultures were not so very dumb, after all.

The outrage that seems to have seized many otherwise placid scientists upon colliding with *Worlds in Collision* has produced a chain of consequences. Some people are quite properly put off by the occasional pomposity of scientists, or are concerned by what they apprehend as the dangers of science and technology, or perhaps merely have difficulty understanding science. They may take some comfort in seeing scientists get their lumps.

In the entire Velikovsky affair, the only aspect worse than the shoddy, ignorant and doctrinaire approach of Velikovsky and many of his supporters was the disgraceful attempt by some who called themselves scientists to suppress his writings. For this, the entire scientific enterprise has suffered. Velikovsky makes no serious claim of objectivity or falsifiability. There is at least nothing hypocritical in his rigid rejection of the immense body of data that contradicts his arguments. But scientists are supposed to know better, to realize that ideas will be judged on their merits if we permit free inquiry and vigorous debate.

To the extent that scientists have not given Velikovsky the reasoned response his work calls for, we have ourselves been responsible for the propagation of Velikovskian confusion. But scientists cannot deal with all areas of borderline science. The thinking, calculations and preparation of this chapter, for example, took badly needed time away from my own research. But it was certainly not boring, and at the very least I had a brush with many an enjoyable legend.

The attempt to rescue old-time religion, in an age which seems desperately to be seeking some religious roots, some cosmic significance for mankind, may or may not be creditable. I think there is much good and much evil in the old-time religions. But I do not understand the need for half-measures. If we are forced to choose between them—and we *decidedly* are not—is the evidence not better for the God of Moses, Jesus and Muhammed than for the comet of Velikovsky?

CHAPTER 8

NORMAN BLOOM, MESSENGER OF GOD

◆◆◆

[The French encyclopedist] Diderot paid a visit to the Russian Court at the invitation of the Empress. He conversed very freely, and gave the younger members of the Court circle a good deal of lively atheism. The Empress was much amused, but some of her councillors suggested that it might be desirable to check these expositions of doctrine. The Empress did not like to put a direct muzzle on her guest's tongue, so the following plot was contrived. Diderot was informed that a learned mathematician was in possession of an algebraical demonstration of the existence of God, and would give it him before all the Court, if he desired to hear it. Diderot gladly consented: though the name of the mathematician is not given, it was Euler. He advanced towards Diderot, and said gravely, and in a tone of perfect conviction: *Monsieur,* $(a + b^n)/n = x$*, donc Dieu existe; répondez!* [Sir, $(a + b^n)/n = x$. Therefore God exists; reply!] Diderot, to whom algebra was Hebrew, was embarrassed and disconcerted; while peals of laughter arose on all sides. He asked permission to return to France at once, which was granted.

AUGUSTUS DE MORGAN,
A Budget of Paradoxes (1872)

THROUGHOUT human history there have been attempts to contrive rational arguments to convince skeptics of the existence of a God or gods. But most theologians have held that the ultimate reality of divine beings is a matter for faith alone and is inaccessible to rational endeavor. St. Anselm argued that since we can imagine a perfect being, he must exist—because he would not be perfect without the added perfection of existence. This so-called ontological argument was more or less promptly attacked on two grounds: (1) *Can* we imagine a completely perfect being? (2) *Is* it obvious that perfection is augmented by existence? To the modern ear such pious arguments seem to be about words and definitions rather than about external reality.

More familiar is the argument from design, an approach that penetrates deeply into issues of fundamental scientific concern. This argument was admirably summarized by David Hume: "Look round the world: contemplate the whole and every part of it; you will find it to be nothing but one great machine, subdivided into an infinite number of lesser machines. . . . All these various machines, even their most minute parts, are adjusted to each other with an accuracy which ravishes into admiration all men who have ever contemplated them. The curious adapting of means to ends, throughout all nature, resembles exactly, though it much exceeds, the production of human contrivance; of human design, thought, wisdom, and intelligence. Since therefore the effects resemble each other, we are led to infer, by all the rules of analogy, that the causes also resemble; and that the Author of Nature is somewhat similar to the mind of man; though possessed of much larger faculties proportioned to the grandure of the work which he has executed."

Hume then goes on to subject this argument, as did Immanuel Kant after him, to a devastating and compelling attack, notwithstanding which the argument from design continued to be immensely popular—as, for example, in the works of William Paley—through the early nineteenth century. A typical passage by Paley goes: "There cannot be a design without a designer; contrivance without a contrivor; order without choice; arrangement without anything capable of arranging; subserviency and relation to a purpose, without that which could intend a purpose; means suitable to an end, and executing their office and accomplishing that end, without the end ever having been contemplated, or the means accommodated to it. Arrangement, disposition of parts, subserviency of means to an end, relation of instruments to a use, imply the presence of intelligence and mind."

It was not until the development of modern science, but most particularly the brilliant formulation of the theory of evolution by natural selection, put forth by Charles Darwin and Alfred Russel Wallace in 1859, that these apparently plausible arguments were fatally undermined.

There can, of course, be no disproof of the existence of God—particularly a sufficiently subtle God. But it is a kindness neither to science nor religion to leave unchallenged inadequate arguments for the existence of God. Moreover, debates on such questions are good fun, and at the very least, hone the mind for useful work. Not much of this sort of disputation is in evidence today, perhaps because new arguments for the existence of God which can be understood at all are exceedingly rare. One recent and modern version of the argument from design was kindly sent to me by its author, perhaps to secure constructive criticism.

NORMAN BLOOM is a contemporary American who incidentally believes himself to be the Second Coming of Jesus Christ. Bloom observes in Scripture and everyday life numerical coincidences which anyone else would consider meaningless. But there are so many such coincidences that, Bloom believes, they can be due only to an unseen intelligence, and the fact that no one else seems to be able to find or appreciate such coincidences convinces Bloom that he has been chosen to reveal God's presence. Bloom has been a fixture at some scientific meetings where he harangues the hurrying, preoccupied crowds moving from session to session. Typical Bloom rhetoric is "And though you reject me, and scorn me, and deny me, YET ALL WILL BE BROUGHT ONLY BY ME. My will will be, because I have formed you out of the nothingness. You are the Creation of My Hands. And I will complete My Creation and Complete My Purpose that I have Purposed from of old. I AM THAT I AM. I AM THE LORD THY GOD IN TRUTH." He is nothing if not modest, and the capitalization conventions are entirely his.

Bloom has issued a fascinating pamphlet, which states: "The complete faculty of Princeton University (including its officers and its deans and the chairmen of the departments listed here) has agreed that it cannot refute, nor show in basic error the proof brought to it, in the book, *The New World* dated Sept. 1974. This faculty acknowledges as of June 1, 1975 that it accepts as a proven truth THE IRREFUTABLE PROOF THAT AN ETERNAL MIND AND HAND HAS SHAPED AND CONTROLLED THE HISTORY OF THE WORLD THROUGH THOUSANDS OF YEARS." A closer reading shows that despite Bloom's distributing his proofs to over a thousand faculty members of Princeton University, and despite his offer of a $1,000 prize for the first individual to refute his proof, there was no response whatever. After six months he concluded that since Princeton did not answer, Princeton believed. Considering the ways of university faculty members, an alternative explanation has occurred to me. In any case, I do not think that the absence of a reply constitutes irrefutable support for Bloom's arguments.

Princeton has apparently not been alone in treating Bloom inhospitably: "Yes, times almost without number, I have been chased by police for bringing you the gift of my writing . . . Is it not so that professors at a university are supposed to have the maturity and judgment and wisdom to be able to read a writing and determine for themselves the value of its contents? Is it that they require THOUGHT CONTROL POLICE to tell them what they should or should not read or think about? But, even at the astronomy department of Harvard University, I have been chased by police for the crime of distributing that New World Lecture, an irrefutable proof that the earth-moon-sun system is shaped by a controlling mind and hand. Yes, and THREATENED WITH IMPRISONMENT, IF I DARE BESMIRCH THE HARVARD CAMPUS WITH MY PRESENCE ONCE MORE. . . . AND THIS IS THE UNIVERSITY THAT HAS UPON ITS SHIELD THE WORD VERITAS: VERITAS: VERITAS:—Truth, Truth, Truth. Ah, what hypocrites and mockers you are!"

The supposed proofs are many and diverse, all involving numerical coincidences which Bloom believes could not be due to chance. Both in style and content, the arguments are reminiscent of Talmudic textual commentary and cabalistic lore of the Jewish Middle Ages: for example, the angular size of the Moon or the Sun as seen from the Earth is half a degree. This is just 1/720 of the circle (360°) of the sky. But $720 = 6! = 6 \times 5 \times 4 \times 3 \times 2 \times 1$. Therefore, God exists. It is an improvement on Euler's proof to Diderot, but the approach is familiar and infiltrates the entire history of religion. In 1658 Gaspar Schott, a Jesuit priest, announced in his *Magia Universalis Naturae et Artis* that the number of degrees of grace of the Virgin Mary is $2^{256} = 2^{2^8} \simeq 1.2 \times 10^{77}$ (which, by and by, is very roughly the number of elementary particles in the universe).

Another Bloomian argument is described as "irrefutable proof that the God of Scripture is he who has shaped and controlled the history of the world through thousands of years." The argument is this: according to Chapters 5 and 11 of Genesis, Abraham was born 1,948 years after Adam, at a time when Abraham's father, Terah, was seventy years old. But the Second Temple was destroyed by the Romans in A.D. 70, and the State of Israel was created in A.D. 1948 Q.E.D. It is hard to escape the impression that there may be a flaw in the argument somewhere. "Irrefutable" is, after all, a fairly strong word. But the argument is a refreshing diversion from St. Anselm.

Bloom's central argument, however, and the one that much of the rest is based upon, is the claimed astronomical coincidence that 235 new moons is, with spectacular accuracy, just as long as nineteen years. Whence: "Look, mankind, I say to you all, in essence you are living in a clock. The clock keeps perfect time, to an accuracy of one second/day! . . . How could such a clock in the heavens come to be without there being some being, who with perception and understanding, who, with a plan and with the power, could form that clock?"

A fair question. To pursue it we must realize that there are several different kinds of years and several different kinds of months in use in astronomy. The sidereal year is the period that the Earth takes to go once around the Sun with respect to the distant stars. It equals 365.2564 days. (The days we will use, as Norman Bloom does, are what astronomers call "mean solar days.") Then there is the tropical year. It is the period for the Earth to make one circuit about the Sun with respect to the seasons, and equals 365.242199 days. The tropical year is different from the sidereal year because of the precession of the equinoxes, the slow toplike movement of the Earth produced by the gravitational forces of the Sun and the Moon on its oblate shape. Finally, there is the so-called anomalistic year of 365.2596 days. It is the interval between two successive closest approaches of the Earth to the Sun, and is different from the sidereal year because of the slow movement of the Earth's elliptical orbit in its own plane, produced by gravitational tugs by the nearby planets.

Likewise, there are several different kinds of months. The word "month," of course, comes from "moon." The sidereal month is the time for the Moon to go once around the earth with respect to the distant stars and equals 27.32166 days. The synodic month, also called a lunation, is the time from new moon to new moon or full moon to full moon. It is 29.530588 days. The synodic month is different from the sidereal month because, in the course of one sidereal revolution of the Moon about the Earth, the Earth-Moon system has together revolved a little bit (about one-thirteenth) of the way around the Sun. Therefore the angle by which the Sun illuminates the Moon has changed from our terrestrial vantage point. Now, the plane of the Moon's orbit around the Earth intersects the plane of the Earth's orbit around the Sun at two places—opposite to each other—called the nodes of the Moon's orbit. A nodical or draconic month is the time for the Moon to move from one node back around again to the same node and equals 27.21220 days. These nodes move, completing one apparent circuit, in 18.6 years because of gravitational tugs, chiefly by the Sun. Finally, there is the anomalistic month of 27.55455 days, which is the time for the Moon to complete one circuit of the Earth with respect to the nearest point in its orbit. A little table on these various definitions of the year and the month is shown below.

KINDS OF YEARS AND MONTHS,
EARTH-MOON SYSTEM

Years	
Sidereal year	365.2564 mean solar days
Tropical year	365.242199 days
Anomalistic year	365.2596 days
Months	
Sidereal month	27.32166 days
Synodic month	29.530588 days
Nodical or draconic month	27.21220 days
Anomalistic month	27.55455 days

Now, Bloom's main proof of the existence of God depends upon choosing one of the sorts of years, multiplying it by 19 and then dividing by one of the sorts of months. Since the sidereal, tropical and anomalistic years are so close together in length, we get sensibly the same answer whichever one we choose. But the same is not true for the months. There are four different kinds of months, and each gives a different answer. If we ask how many *synodic* months there are in nineteen sidereal years, we find the answer to be 253.00621, as advertised; and it is the closeness of this result to a whole number that is the fundamental coincidence of Bloom's thesis. Bloom, of course, believes it to be no coincidence.

But if we were to ask instead how many *sidereal* months there are in nineteen sidereal years we would find the answer to be 254.00622; for nodical months, 255.02795; and for anomalistic months, 251.85937. It is certainly true that the synodic month is the one most strikingly apparent to a naked-eye observer, but I nevertheless have the impression that one could construct equally elaborate theological speculations on 252, 254, or 255 as on 235.

We must now ask where the number 19 comes from in this argument. Its only justification is David's lovely Nineteenth Psalm, which begins: "The heavens declare the glory of God, and the firmament sheweth his handiwork. Day unto day uttereth speech, and night unto night sheweth knowledge." This seems quite an appropriate quotation from which to find a hint of an astronomical proof for the existence of God. But the argument assumes what it intends to prove. The argument is also not unique. Consider, for example, the Eleventh Psalm, also written by David. In it we find the following words, which may equally well bear on this question: "The Lord is in his holy temple, the Lord's throne is in heaven: his eyes behold, his eyelids try, the children of men," which is followed in the following Psalm with "the children of men . . . speak vanity." Now, if we ask how many synodic months there are in eleven sidereal years (or 4017.8204 mean solar days), we find the answer to be 136.05623. Thus, just as there seems to be a connection between nineteen years and 235 new moons, there is a connection between eleven years and 136 new moons. Moreover, the famous British astronomer Sir Arthur Stanley Eddington believed that all of physics could be derived from the number 136.

(I once suggested to Bloom that with the foregoing information and just a little intellectual fortitude it should be possible as well to reconstruct all of Bosnian history.)

One numerical coincidence of this sort, which *is* of deep significance, was well known to the Babylonians, contemporaries of the ancient Hebrews. It is called the Saros. It is the period between two successive similar cycles of eclipses. In a solar eclipse the Moon, which appears from the Earth just as large (1/2°) as the Sun, must pass in front of it. For a lunar eclipse, the Earth's shadow in space must intercept the Moon. For either kind of eclipse to occur, the Moon must, first of all, be either new or full—so that the Earth, the Moon and the Sun are in a straight line. Therefore the synodic month is obviously involved in the periodicity of eclipses. But for an eclipse to occur, the Moon must also be near one of the nodes of its orbit. Therefore the nodical month is involved. It turns out that 233 synodic months is equal to 241.9989 (or very close to 242) nodical months. This is the equivalent of a little over eighteen years and ten or eleven days (depending on the number of intervening leap days), and comprises the Saros. Coincidence?

Similar numerical coincidences are in fact common throughout the solar system. The ratio of spin period to orbital period on Mercury is 3 to 2. Venus manages to turn the same face to the Earth at its closest approach on each of its revolutions around the Sun. A particle in the gap between the two principal rings of Saturn, called the Cassini Division, would orbit Saturn in a period just half that of Mimas, its second satellite. Likewise, in the asteroid belt there are empty regions, known as the Kirkwood Gaps, which correspond to nonexistent asteroids with periods half that of Jupiter, one-third, two-fifths, three-fifths, and so on.

None of these numerical coincidences proves the existence of God—or if it does, the argument is subtle, because these effects are due to resonances. For example, an asteroid that strays into one of the Kirkwood Gaps experiences a periodic gravitational pumping by

Jupiter. Every two times around the Sun for the asteroid, Jupiter makes exactly one circuit. There it is, tugging away at the same point in the asteroid's orbit every revolution. Soon the asteroid is persuaded to vacate the gap. Such incommensurable ratios of whole numbers are a general consequence of gravitational resonance in the solar system. It is a kind of perturbational natural selection. Given enough time—and time is what the solar system has a great deal of—such resonances will arise inevitably.

That the general result of planetary perturbations is stable resonances and not catastrophic collisions was first shown from Newtonian gravitational theory by Pierre Simon, Marquis de Laplace, who described the solar system as "a great pendulum of eternity, which beats ages as a pendulum beats seconds." Now, the elegance and simplicity of Newtonian gravitation might be used as an argument for the existence of God. We could imagine universes with other gravitational laws and much more chaotic planetary interactions. But in many of those universes we would not have evolved—precisely because of the chaos. Such gravitational resonances do not prove the existence of God, but if he does exist, they show, in the words of Einstein, that, while he may be subtle, he is not malicious.

BLOOM CONTINUES his work. He has, for example, demonstrated the preordination of the United States of America by the prominence of the number 13 in major league baseball scores on July 4, 1976. He has accepted my challenge and made an interesting attempt to derive some of Bosnian history from numerology—at least the assassination of Archduke Ferdinand at Sarajevo, the event that precipitated World War I. One of his arguments involves the date on which Sir Arthur Stanley Eddington presented a talk on his mystical number 136 at Cornell University, where I teach. And he has even performed some numerical manipulations using my birth date to demonstrate that I also am part of the cosmic plan. These and similar cases convince me that Bloom can prove anything.

Norman Bloom is, in fact, a kind of genius. If enough independent phenomena are studied and correlations sought, some will of course be found. If we know only the coincidences and not the enormous effort and many unsuccessful trials that preceded their discovery, we might believe that an important finding has been made. Actually, it is only what statisticians call "the fallacy of the enumeration of favorable circumstances." But to find as many coincidences as Norman Bloom has requires great skill and dedication. It is in a way a forlorn and perhaps even hopeless objective—to demonstrate the existence of God by numerical coincidences to an uninterested, to say nothing of a mathematically unenlightened public. It is easy to imagine the contributions Bloom's talents might have made in another field. But there is something a little glorious, I find, in his fierce dedication and very considerable arithmetic intuition. It is a combination of talents which is, one might almost say, God-given.

CHAPTER 9

SCIENCE FICTION—A PERSONAL VIEW

◆◆◆

The poet's eye, in a fine frenzy rolling,
Doth glance from heaven to earth, from earth to heaven;
And as imagination bodies forth
The forms of things unknown, the poet's pen
Turns them to shapes, and gives to airy nothing
A local habitation and a name.

WILLIAM SHAKESPEARE,
A Midsummer Night's Dream, Act V, Scene 1

BY THE TIME I was ten I had decided—in almost total ignorance of the difficulty of the problem—that the universe was full up. There were too many places for this to be the only inhabited planet. And judging from the variety of life on Earth (trees looked pretty different from most of my friends), I figured life elsewhere would look very strange. I tried hard to imagine what that life would be like, but despite my best efforts I always produced a kind of terrestrial chimaera, a blend of existing plants and animals.

About this time a friend introduced me to the Mars

novels of Edgar Rice Burroughs. I had not thought much about Mars before, but here, presented before me in the adventures of John Carter, was an inhabited extraterrestrial world breathtakingly fleshed out: ancient sea bottoms, great canal-pumping stations and a variety of beings, some of them exotic. There were, for example, the eight-legged beasts of burden, the thoats.

These novels were exhilarating to read. At first. Then slowly doubts began to gnaw. The plot surprise in the first John Carter novel I read hinged on him forgetting that the year is longer on Mars than on Earth. But it seemed to me that if you go to another planet, one of the first things you check into is the length of the day and the year. (Incidentally, I can recall no mention by Carter of the remarkable fact that the Martian day is almost as long as the terrestrial day. It was as if he *expected* the familiar features of his home planet somewhere else.) Then there were incidental remarks made which were at first stunning but on sober reflection disappointing. For example, Burroughs casually comments that on Mars there are two more primary colors than on Earth. I spent many long minutes with my eyes tightly closed, fiercely concentrating on a new primary color. But it would always be a murky brown or a plum. How could there be another primary color on Mars, much less two? What *was* a primary color? Was it something to do with physics or something to do with physiology? I decided that Burroughs might not have known what he was talking about, but he certainly made his readers think. And in those many chapters where there was not much to think about, there were satisfyingly malignant enemies and rousing swordsmanship—more than enough to maintain the interest of a citybound ten-year-old in a Brooklyn summer.

A year later, by sheerest accident, I stumbled across a magazine called *Astounding Science Fiction* in the neighborhood candy store. A glance at the cover and a quick riffle through the interior showed me it was what I had been looking for. With some effort I managed to scrape together the purchase price, opened it at random, sat down on a bench not twenty feet from the candy store and read my first modern science-fiction short story, "Pete Can Fix It," by Raymond F. Jones, a gentle time-travel story of post–nuclear-war holocaust. I knew about the atom bomb—I remember an excited friend explaining to me that it was made of atoms—but this was the first I had seen about the social implications of the development of nuclear weapons. It got you thinking. The little device, though, that Pete the garage mechanic put on automobiles so passers-by might make brief cautionary trips into the wasteland of the future—what was that little device? How was it made? How could you get into the future and then come back? If Raymond F. Jones knew, he wasn't telling.

I found I was hooked. Each month I eagerly awaited the arrival of *Astounding*. I read Jules Verne and H. G. Wells, read from cover to cover the first two science-fiction anthologies that I was able to find, made scorecards, similar to those I was fond of making for baseball, on the quality of the stories I read. Many of the stories ranked high in asking interesting questions but low in answering them.

There is still a part of me that is ten years old. But by and large I'm older. My critical faculties and perhaps even my literary tastes have improved. In rereading L. Ron Hubbard's *The End Is Not Yet,* which I had first read at age fourteen, I was so amazed at how much worse it was than I had remembered that I seriously considered the possibility that there were two novels of the same name and by the same author but of vastly differing quality. I can no longer manage credulous acceptance as well as I used to. In Larry Niven's *Neutron Star* the plot hinges on the astonishing tidal forces exerted by a strong gravitational field. But we are asked to believe that hundreds or thousands of years from now, at a time of casual interstellar spaceflight, such tidal forces have been forgotten. We are asked to believe that the first probe of a neutron star is done by a manned rather than by an unmanned spacecraft. We are asked too much. In a novel of ideas, the ideas have to work.

I had the same kind of disquieting feelings many years

earlier on reading Verne's description that weightlessness on a lunar voyage occurred only at the point in space where the Earth's and the Moon's gravitational pulls canceled, and in Wells's invention of the antigravity mineral cavorite: Why should a vein of cavorite still be on Earth? Shouldn't it have flung itself into space long ago? In Douglas Trumbull's technically proficient science-fiction film *Silent Running,* the trees in vast closed spaceborne ecological systems are dying. After weeks of painstaking study and agonizing searches through botany texts, the solution is found: plants, it turns out, need sunlight. Trumbull's characters are able to build interplanetary cities but have forgotten the inverse square law. I was willing to overlook the portrayal of the rings of Saturn as pastel-colored gases, but not this.

I have the same trouble with *Star Trek,* which I know has a wide following and which some thoughtful friends tell me I should view allegorically and not literally. But when astronauts from Earth set down on some far-distant planet and find the human beings there in the midst of a conflict between two nuclear superpowers—which call themselves the Yangs and the Coms, or their phonetic equivalents—the suspension of disbelief crumbles. In a global terrestrial society centuries in the future, the ship's officers are embarrassingly Anglo-American. Only two of twelve or fifteen interstellar vessels are given non-English names, *Kongo* and *Potemkin. (Potemkin* and not *Aurora?)* And the idea of a successful cross between a "Vulcan" and a terrestrial simply ignores what we know of molecular biology. (As I have remarked elsewhere, such a cross is about as likely as the successful mating of a man and a petunia.) According to Harlan Ellison, even such sedate biological novelties as Mr. Spock's pointy ears and permanently querulous eyebrows were considered by network executives far too daring; such enormous differences between Vulcans and humans would only confuse the audience, they thought, and a move was made to have all physiologically distinguishing Vulcanian features effaced. I have similar problems with films in which familiar creatures, slightly changed—spiders thirty feet tall—are menacing the cities of the Earth: since insects and arachnids breathe by diffusion, such marauders would asphyxiate before they could savage their first city.

I believe that the same thirst for wonder is inside me that was there when I was ten. But I have learned since then a little bit about how the world is really put together. I find that science fiction has led me to science. I find science more subtle, more intricate and more awesome than much of science fiction. Think of some of the scientific findings of the last few decades: that Mars is covered with ancient dry rivers; that apes can learn languages of many hundreds of words, understand abstract concepts and construct new grammatical usages; that there are particles that pass effortlessly through the entire Earth so that we see as many of them coming up through our feet as down from the sky; that in the constellation Cygnus there is a double star, one of whose components has such a high gravitational acceleration that light cannot escape from it: it may be blazing with radiation on the inside but it is invisible from the outside. In the face of all this, many of the standard ideas of science fiction seem to me to pale by comparison. I see the relative absence of these things and the distortions of scientific thinking often encountered in science fiction as terrible wasted opportunities. Real science is as amenable to exciting and engrossing fiction as fake science, and I think it is important to exploit every opportunity to convey scientific ideas in a civilization which is both based upon science and does almost nothing to ensure that science is understood.

But the best of science fiction remains very good indeed. There are stories so tautly constructed, so rich in accommodating details of an unfamiliar society that they sweep me along before I even have a chance to be critical. Such stories include Robert Heinlein's *The Door into Summer*, Alfred Bester's *The Stars My Destination* and *The Demolished Man*, Jack Finney's *Time and Again*, Frank Herbert's *Dune* and Walter M. Miller's *A Canticle for Leibowitz*. You can ruminate over the ideas in these books. Heinlein's asides on the feasibility and social utility of household robots wear exceedingly well over the years. The insights into terrestrial ecology provided by hypothetical extraterrestrial ecologies as in *Dune* perform, I think, an important social service. *He Who Shrank,* by Harry Hasse, presents an entrancing cosmological speculation which is being seriously revived today, the idea of an infinite regress of universes—in which each of our elementary particles is a universe one level down, and in which we are an elementary particle in the next universe up.

A rare few science-fiction novels combine extraordinarily well a deep human sensitivity with a standard science-fiction theme. I am thinking, for example, of Algis Budrys' *Rogue Moon,* and of many of the works of Ray Bradbury and Theodore Sturgeon—for example, the latter's *To Here and the Easel,* a stunning portrayal of schizophrenia as perceived from the inside, as well as a provocative introduction to Ariosto's *Orlando Furioso.*

There was once a subtle science-fiction story by the astronomer Robert S. Richardson on the continuous-creation origin of cosmic rays. Isaac Asimov's story *Breathes There a Man* provided a poignant insight into the emotional stress and sense of isolation of some of the best theoretical scientists. Arthur C. Clarke's *The Nine Billion Names of God* introduced many Western readers to an intriguing speculation in Oriental religions.

One of the great benefits of science fiction is that it can convey bits and pieces, hints and phrases, of knowledge unknown or inaccessible to the reader. Heinlein's *And He Built a Crooked House* was for many readers probably the first introduction they had ever encountered to four-dimensional geometry that held any promise of being comprehensible. One science-fiction work actually presents the mathematics of Einstein's last attempt at a unified field theory; another presents an important equation in population genetics. Asimov's robots were "positronic," because the positron had recently been discovered. Asimov never provided any explanation of how positrons run robots, but his readers had now heard of positrons. Jack Williamson's rhodomagnetic robots were run off ruthenium, rhodium and palladium, the next Group VIII metals after iron, nickel and cobalt in the periodic table. An analogue with ferromagnetism was suggested. I suppose that there are science-fiction robots today that are quark-ish or charming and will provide some brief verbal entrée into the excitement of contemporary elementary particle physics. L. Sprague de Camp's *Lest Darkness Fall* is an excellent introduction to Rome at the time of the Gothic invasion, and Asimov's *Foundation* series, although this is not explained in the books, offers a very useful summary of some of the dynamics of the far-flung imperial Roman Empire. Time-travel stories—for example, the three remarkable efforts by Heinlein, *All You Zombies, By His Bootstraps* and *The Door into Summer*—force the reader into contemplations of the nature of causality and the arrow of time. They are books you ponder over as the water is running out of the bathtub or as you walk through the woods in an early winter snowfall.

Another great value of modern science fiction is some of the art forms it elicits. A fuzzy imagining in the mind's eye of what the surface of another planet might look like is one thing, but examining a meticulous painting of the same scene by Chesley Bonestell in his prime is quite another. The sense of astronomical wonder is splendidly conveyed by the best of such contemporary artists—Don Davis, Jon Lomberg, Rick Sternbach, Robert McCall. And in the verse of Diane Ackerman can be glimpsed the prospect of a mature astronomical poetry, fully conversant with standard science-fiction themes.

Science-fiction ideas are widespread today in somewhat different guises. We have science-fiction writers such as Isaac Asimov and Arthur C. Clarke providing cogent and brilliant summaries in nonfictional form of many aspects of science and society. Some contemporary scientists are introduced to a vaster public by science fiction. For example, in the thoughtful novel *The Listeners,* by James Gunn, we find the following comment made fifty years from now about my colleague, the astronomer Frank Drake: "Drake! What did he know?" A great deal, it turns out. We also find straight science fiction disguised as fact in a vast proliferation of pseudoscientific writings, belief systems and organizations.

One science-fiction writer, L. Ron Hubbard, has founded a successful cult called Scientology—invented, according to one account, overnight on a bet that he could do as well as Freud, invent a religion and make money from it. Classic science-fiction ideas are now institutionalized in unidentified flying objects and ancient-astronaut belief systems—although I have difficulty not concluding that Stanley Weinbaum (in *The Valley of Dreams*) did it better, as well as earlier, than Erich von Däniken. R. De Witt Miller in *Within the Pyramid* manages to anticipate both von Däniken and Velikovsky, and to provide a more coherent hypothesis on the supposed extraterrestrial origin of pyramids than can be found in all the writings on ancient astronauts and pyramidology. In *Wine of the Dreamers,* by John D. MacDonald (a science-fiction author now transmogrified into one of the most interesting contemporary writers of detective fiction), we find the sentence "and there are traces, in Earth mythology . . . of great ships and chariots that crossed the sky." The story *Farewell to the Master,* by Harry Bates, was converted into a motion picture, *The Day the Earth Stood Still* (which abandoned the essential plot element, that on the extraterrestrial spacecraft it was the robot and not the human who was in command). The movie, with its depiction of a flying saucer buzzing Washington, is thought by some sober investigators to have played a role in the 1952 Washington, D.C., UFO "flap" which followed closely the release of the motion picture. Many popular novels today of the espionage variety, in the shallowness of their characterizations and the gimmickry of their plots, are virtually indistinguishable from pulp science fiction of the '30s and '40s.

THE INTERWEAVING of science and science fiction sometimes produces curious results. It is not always clear whether life imitates art or vice versa. For example,

Kurt Vonnegut, Jr., has written a superb epistemological novel, *The Sirens of Titan,* in which a not altogether inclement environment is postulated on Saturn's largest moon. When in the last few years some planetary scientists, myself among them, presented evidence that Titan has a dense atmosphere and perhaps higher temperatures than expected, many people commented to me on the prescience of Kurt Vonnegut. But Vonnegut was a physics major at Cornell University and naturally knowledgeable about the latest findings in astronomy. (Many of the best science-fiction writers have science or engineering backgrounds; for example, Poul Anderson, Isaac Asimov, Arthur C. Clarke, Hal Clement and Robert Heinlein.) In 1944, an atmosphere of methane was discovered on Titan, the first satellite in the solar system known to have an atmosphere. In this, as in many similar cases, art imitates life.

The trouble has been that our understanding of the other planets has been changing faster than the science-fiction representations of them. A clement twilight zone on a synchronously rotating Mercury, a swamp-and-jungle Venus and a canal-infested Mars, while all classic science-fiction devices, are all based upon earlier misapprehensions by planetary astronomers. The erroneous ideas were faithfully transcribed into science-fiction stories, which were then read by many of the youngsters who were to become the next generation of planetary astronomers—thereby simultaneously capturing the interest of the youngsters and making it more difficult to correct the misapprehensions of the oldsters. But as our knowledge of the planets has changed, the environments in the corresponding science-fiction stories have also changed. It is quite rare to find a science-fiction story written today that involves algae farms on the surface of Venus. (Incidentally, the UFO-contact mythologizers are slower to change, and we can still find accounts of flying saucers from a Venus populated by beautiful human beings in long white robes inhabiting a kind of Cytherean Garden of Eden. The 900° Fahrenheit temperatures of Venus give us one way of checking such stories.) Likewise, the idea of a "space warp" is a hoary science-fiction standby but it did not arise in science fiction. It arose from Einstein's General Theory of Relativity.

The connection between science-fiction depictions of Mars and the actual exploration of Mars is so close that, subsequent to the Mariner 9 mission to Mars, we were able to name a few Martian craters after deceased science-fiction personalities. (See Chapter 11.) Thus there are on Mars craters named after H. G. Wells, Edgar Rice Burroughs, Stanley Weinbaum and John W. Campbell, Jr. These names have been officially approved by the International Astronomical Union. No doubt other science-fiction personalities will be added soon after they die.

THE GREAT INTEREST of youngsters in science fiction is reflected in films, television programs, comic books and a demand for science-fiction courses in high schools and colleges. My experience is that such courses can be fine educational experiences or disasters, depending on how they are done. Courses in which the readings are selected by the students provide no opportunity for the students to read what they have not already read. Courses in which there is no attempt to extend the science-fiction plot line to encompass the appropriate science miss a great educational opportunity. But properly planned science-fiction courses, in which science or politics is an integral component, would seem to me to have a long and useful life in school curricula.

The greatest human significance of science fiction may be as experiments on the future, as explorations of alternative destinies, as attempts to minimize future shock. This is part of the reason that science fiction has so wide an appeal among young people: it is *they* who will live in the future. It is my firm view that no society on Earth today is well adapted to the Earth of one or two hundred years from now (if we are wise enough or lucky enough to survive that long). We desperately need an exploration of alternative futures, both experimental and conceptual. The novels and stories of Eric Frank Russell were very much to this point. In them, we were able to see conceivable alternative economic systems or the great efficiency of a unified passive resistance to an occupying power. In modern science fiction, useful suggestions can also be found for making a revolution in a computerized technological society, as in Heinlein's *The Moon Is a Harsh Mistress.*

Such ideas, when encountered young, can influence adult behavior. Many scientists deeply involved in the exploration of the solar system (myself among them) were first turned in that direction by science fiction. And the fact that some of that science fiction was not of the highest quality is irrelevant. Ten-year-olds do not read the scientific literature.

I do not know if time travel into the past is possible. The causality problems it would imply make me very skeptical. But there are those who are thinking about it. What are called closed time-like lines—routes in space-time permitting unrestricted time travel—appear in some solutions to the general relativistic field equations. A recent claim, perhaps mistaken, is that closed time-like lines appear in the vicinity of a large, rapidly rotating cylinder. I wonder to what extent general-relativists working on such problems have been influenced by science fiction. Likewise, science-fiction encounters with alternative cultural features may play an important role in actualizing fundamental social change.

In all the history of the world there has never before been a time in which so many significant changes have occurred. Accommodation to change, the thoughtful pursuit of alternative futures are keys to the survival of civilization and perhaps of the human species. Ours is the first generation that has grown up with science-fiction ideas. I know many young people who will of course be interested but in no way astounded if we receive a message from an extraterrestrial civilization. They have already accommodated to that future. I think it is no exaggeration to say that if we survive, science fiction will have made a vital contribution to the continuation and evolution of our civilization.

PART III

OUR NEIGHBORHOOD IN SPACE

CHAPTER 10

THE SUN'S FAMILY

◆◆◆

> Like a shower of stars the worlds whirl, borne along by the winds of heaven, and are carried down through immensity; suns, earths, satellites, comets, shooting stars, humanities, cradles, graves, atoms of the infinite, seconds of eternity, perpetually transform beings and things.
>
> CAMILLE FLAMMARION,
> *Popular Astronomy*, translated by J. E. Gore
> (New York, D. Appleton & Company, 1894)

IMAGINE THE EARTH scrutinized by some very careful and extremely patient extraterrestrial observer: 4.6 billion years ago the planet is observed to complete its condensation out of interstellar gas and dust, the final planetesimals falling in to make the Earth produce enormous impact craters; the planet heats internally from the gravitational potential energy of accretion and from radioactive decay, differentiating the liquid iron core from the silicate mantle and crust; hydrogen-rich gases and condensible water are released from the interior of the planet to the surface; a rather humdrum cosmic organic chemistry yields complex molecules, which lead to extremely simple self-replicating molecular systems—the first terrestrial organisms; as the supply of impacting interplanetary boulders dwindles, running water, mountain building and other geological processes wipe out the scars attendant to the Earth's origin; a vast planetary convection engine is established which carries mantle material up at the ocean floors and subducts it down at the continental margins, the collision of the moving plates producing the great folded mountain chains and the general configuration of land and ocean, glaciated and tropical terrain varies continuously. Meanwhile, natural selection extracts out from a wide range of alternatives those varieties of self-replicating molecular systems best suited to the changing environments; plants evolve that use visible light to break down water into hydrogen and oxygen, and the hydrogen escapes to space, changing the chemical composition of the atmosphere from reducing to oxidizing; organisms of fair complexity and middling intelligence eventually arise.

Yet in all the 4.6 billion years our hypothetical observer is struck by the isolation of the Earth. It receives sunlight and cosmic rays—both important for biology—and occasional impact of interplanetary debris. But nothing in all those eons of time leaves the planet. And then the planet suddenly begins to fire tiny dispersules throughout the inner solar system, first in orbit around the Earth, then to the planet's blasted and lifeless natural satellite, the Moon. Six capsules—small, but larger than the rest—set down on the Moon, and from each, two tiny bipeds can be discerned, briefly exploring their surroundings and then hotfooting it back to the Earth, having extended tentatively a toe into the cosmic ocean. Eleven little spacecraft enter the atmosphere of Venus, a searing hellhole of a world, and six of them survive some tens of minutes on the surface before being fried. Eight spacecraft are sent to Mars. Three successfully orbit the planet for years; another flies past Venus to encounter Mercury, on a trajectory obviously chosen intentionally to pass by the innermost planet many times. Four others successfully traverse the asteroid belt, fly close to Jupiter and are there ejected by the gravity of the largest planet into interstellar space. It is clear that something interesting is happening lately on the planet Earth.

If the 4.6 billion years of the Earth history were compressed into a single year, this flurry of space exploration would have occupied the last tenth of a second, and the fundamental changes in attitude and knowledge responsible for this remarkable transformation would fill only the last few seconds. The seventeenth century saw the first widespread application of simple lenses and mirrors for astronomical purposes. With the first astronomical telescope Galileo was astounded and delighted to see Venus as a crescent, and the mountains and the craters of the Moon. Johannes Kepler thought that the craters were constructions of intelligent beings inhabiting that world. But the seventeenth-century Dutch physicist Christianus Huygens disagreed. He suggested that the effort involved in constructing the lunar craters would be unreasonably great, and also thought that he could see alternative explanations for these circular depressions.

Huygens exemplified the synthesis of advancing technology, experimental skills, a reasonable, hard-nosed and skeptical mind, and an openness to new ideas. He was the first to suggest that we are looking at atmosphere and clouds on Venus; the first to understand something of the true nature of the rings of Saturn (which had seemed to Galileo as two "ears" enveloping the planet); the first to draw a picture of a recognizable marking on the Martian surface (Syrtis Major); and the second, after Robert Hooke, to draw the Great Red Spot of Jupiter. These last two observations are still of scientific importance because they establish the permanence at least for three centuries of these features. Huygens was of course not a thoroughly modern astronomer. He could not entirely escape the fashions of belief of his time. For example, he presented a curious argument from which we could deduce the presence of hemp on Jupiter: Galileo had observed that Jupiter has four moons. Huygens asked a question few modern planetary astronomers would ask: *Why* does Jupiter have four moons? An insight into this question, he thought, could be garnered by asking the same question of the Earth's single moon, whose function, apart from giving a little light at night and raising the tides, was to provide a navigational aid to mariners. If Jupiter has four moons, there must be many mariners on that planet. But mariners imply boats; boats imply sails; sails imply ropes; and, I suppose, ropes imply hemp. I wonder how many of our present highly prized scientific arguments will seem equally suspect from the vantage point of three centuries.

A useful index of our knowledge about a planet is the number of bits of information necessary to characterize our understanding of its surface. We can think of this as the number of black and white dots in the equivalent of a newspaper wirephoto which, held at arm's length, would summarize all existing imagery. Back in Huygens' day, about ten bits of information, all obtained by brief glimpses through telescopes, would have covered our knowledge of the surface of Mars. By the time of the close approach of Mars to Earth in the year 1877, this number had risen to perhaps a few thousand, if we exclude a large amount of erroneous information—for example, drawings of the "canals," which we now know to be entirely illusory. With further visual observations and the development of ground-based astronomical photography, the amount of information grew slowly until a dramatic upturn in the curve occurred, corresponding to the advent of space-vehicle exploration of the planet.

The twenty photographs obtained in 1965 by the Mariner 4 fly-by comprised five million bits of information, roughly comparable to all previous photographic knowledge about the planet. The coverage was still only a tiny fraction of the planet. The dual fly-by mission, Mariner 6 and 7 in 1969, increased this number by a factor of 100, and the Mariner 9 orbiter in 1971 and 1972 increased it by another factor of 100. The Mariner 9 photographic results from Mars correspond roughly to 10,000 times the total previous photographic knowledge of Mars obtained over the history of mankind. Comparable improvements apply to the infrared and ultraviolet spectroscopic data obtained by Mariner 9, compared with the best previous ground-based data.

Going hand in hand with the improvement in the quantity of our information is the spectacular improvement in its quality. Prior to Mariner 4, the smallest feature reliably detected on the surface of Mars was several hundred kilometers across. After Mariner 9, several percent of the planet had been viewed at an effective resolution of 100 meters, an improvement in resolution of a factor of 1,000 in the last ten years, and a factor of 10,000 since Huygens' time. Still further improvements were provided by Viking. It is only because of this improvement in resolution that we today know of vast volcanoes, polar laminae, sinuous tributaried channels, great rift valleys, dune fields, crater-associated dust streaks, and many other features, instructive and mysterious, of the Martian environment.

Both resolution and coverage are required to understand a newly explored planet. For example, even with their superior resolution, by an unlucky coincidence the Mariner 4, 6 and 7 spacecraft observed the old, cratered and relatively uninteresting part of Mars and gave no hint of the young and geologically active third of the planet revealed by Mariner 9.

LIFE ON EARTH is wholly undetectable by orbital photography until about 100-meter resolution is achieved, at which point the urban and agricultural geometrizing of our technological civilization becomes strikingly evident. Had there been a civilization on Mars of comparable extent and level of development, it would not have been detected photographically until the Mariner 9 and Viking missions. There is no reason to expect such civilizations on the nearby planets, but the comparison strikingly illustrates that we are just beginning an adequate reconnaissance of neighboring worlds.

THERE IS NO question that astonishments and delights await us as both resolution and coverage are dramatically improved in photography, and comparable improvements are secured in spectroscopic and other methods.

The largest professional organization of planetary scientists in the world is the Division for Planetary Sciences of the American Astronomical Society. The vigor of this burgeoning science is apparent in the meetings of the society. In the 1975 annual meeting, for example, there were announcements of the discovery of water vapor in the atmosphere of Jupiter, ethane on Saturn, possible hydrocarbons on the asteroid Vesta, an atmospheric pressure approaching that of the Earth on the Saturnian moon Titan, decameter-wavelength radio bursts from Saturn, the radar detection of the Jovian moon Ganymede, the elaboration of the radio emission spectrum of the Jovian moon Callisto, to say nothing of the spectacular views of Mercury and Jupiter (and their magnetospheres) presented by the Mariner 10 and Pioneer 11 experiments. Comparable advances were reported in subsequent meetings.

In all the flurry and excitement of recent discoveries, no general view of the origin and evolution of the planets has yet emerged, but the subject is now very rich in provocative hints and clever surmises. It is becoming clear that the study of any planet illuminates our knowledge of the rest, and if we are to understand Earth thoroughly, we must have a comprehensive knowledge of the other planets. For example, one now fashionable suggestion, which I first proposed in 1960, is that the high temperatures on the surface of Venus are due to a runaway greenhouse effect in which water and carbon dioxide in a planetary atmosphere impede the emission of thermal infrared radiation from the surface to space; the surface temperature then rises to achieve equilibrium between the visible sunlight arriving at the surface and the infrared radiation leaving it; this higher surface temperature results in a higher vapor pressure of the greenhouse gases, carbon dioxide and water; and so on, until all the carbon dioxide and water vapor is in the vapor phase, producing a planet with high atmospheric pressure and high surface temperature.

Now, the reason that Venus has such an atmosphere and Earth does not seems to be a relatively small increment of sunlight. Were the Sun to grow brighter or Earth's surface and clouds to grow darker, could Earth become a replica of the classical vision of Hell? Venus may be a cautionary tale for our technical civilization, which has the capability to alter profoundly the environment of Earth.

Despite the expectation of almost all planetary scientists, Mars turns out to be covered with thousands of sinuous tributaried channels probably several billion years old. Whether formed by running water or running CO_2, many such channels probably could not be carved under present atmospheric conditions; they require much higher pressures and probably higher polar temperatures. Thus the channels—as well as the polar laminated terrain on Mars—may bear witness to at least one, and perhaps many, previous epochs of much more clement conditions, implying major climatic variations during the history of the planet. We do not know if such variations are internally or externally caused. If internally, it will be of interest to see whether the Earth might, through the activities of man, experience a Martian degree of climatic excursions—something much greater than the Earth seems to have experienced at least recently. If the Martian climatic variations are externally produced—for example, by variations in solar luminosity—then a correlation of Martian and terrestrial paleoclimatology would appear extremely promising.

Mariner 9 arrived at Mars in the midst of a great global dust storm, and the Mariner 9 data permit an observational test of whether such storms heat or cool a planetary surface. Any theory with pretensions to predicting the climatic consequences of increased aerosols in the Earth's atmosphere had better be able to provide the correct answer for the global dust storm observed by Mariner 9. Drawing upon our Mariner 9 experience, James Pollack of NASA Ames Research Center, Brian Toon of Cornell and I have calculated the effects of single and multiple volcanic explosions on the Earth's climate and have been able to reproduce, within experimental error, the observed climatic effects after major explosions on our planet. The perspective of planetary astronomy, which permits us to view a planet as a whole, seems to be very good training for studies of the Earth. As another example of this feedback from planetary studies on terrestrial observations, one of the major groups studying the effect on the Earth's ozonosphere of the use of halocarbon propellants from aerosol cans is headed by M. B. McElroy at Harvard University—a group that cut its teeth for this problem on the aeronomy of the atmosphere of Venus.

We now know from space-vehicle observations something of the surface density of impact craters of different sizes for Mercury, the Moon, Mars and its satellites; radar studies are beginning to provide such information for Venus, and although it is heavily eroded by running water and tectonic activity, we have some information about craters on the surface of the Earth. If the population of objects producing such impacts were the same for all these planets, it might then be possible to establish both an absolute and a relative chronology of cratered surfaces. But we do not yet know whether the populations of impacting objects are common—all derived from the asteroid belt, for example—or local; for example, the sweeping up of rings of debris involved in the final stages of planetary accretion.

The heavily cratered lunar highlands speak to us of an early epoch in the history of the solar system when cratering was much more common than it is today; the present population of interplanetary debris fails by a large factor to account for the abundance of the highland craters. On the other hand, the lunar maria have a much lower crater abundance, which can be explained by the present population of interplanetary debris, largely asteroids and possibly dead comets. It is possible to determine, for planetary surfaces that are not so heavily cratered, something of the absolute age, a great deal about the relative age, and in some cases, even something about the distribution of sizes in the population of objects that produced the craters. On Mars, for example, we find the flanks of the large volcanic mountains are almost free of impact craters, implying their comparative youth; they were not around long enough to accumulate very much in the way of impact scars. This is the basis for the contention that volcanoes on Mars are a comparatively recent phenomenon.

The ultimate objective of comparative planetology is, I suppose, something like a vast computer program into which we put a few input parameters—perhaps the initial mass, composition, angular momentum and population of neighboring impacting objects—and out comes the time evolution of the planet. We are very far from having such a deep understanding of planetary evolution at the present time, but we are much closer than would have been thought possible only a few decades ago.

Every new set of discoveries raises a host of questions which we were never before wise enough even to ask. I will mention just a few of them. It is now becoming possible to compare the compositions of asteroids with the compositions of meteorites on Earth (see Chapter 15). Asteroids seem to divide neatly into silicate-rich and organic-matter–rich objects. One immediate consequence appears to be that the asteroid Ceres is apparently undifferentiated, while the less massive asteroid Vesta is differentiated. But our present understanding is that planetary differentiation occurs above a certain critical mass. Could Vesta be the remnant of a much larger parent body now gone from the solar system? The initial radar glimpse of the craters of Venus shows them to be extremely shallow. Yet there is no liquid water to erode the Venus surface, and the lower atmosphere of Venus seems to be so slow-moving that dust may not be able to fill the craters. Could the source of the filling of the craters of Venus be a slow molasses-like collapse of a very slightly molten surface?

The most popular theory on the generation of planetary magnetic fields invokes rotation-driven convection currents in a conducting planetary core. Mercury, which rotates once every fifty-nine days, was expected in this scheme to have no detectable magnetic field. Yet such a field is manifestly there, and a serious reappraisal of theories of planetary magnetism is in order. Only Saturn and Uranus have rings. Why? There is on Mars an exquisite array of longitudinal sand dunes nestling against the interior ramparts of a large eroded crater. There is in the Great Sand Dunes National Monument near Alamosa, Colorado, a very similar set of sand dunes nestling in the curve of the Sangre de Cristo mountains. The Martian and the terrestrial sand dunes have the same total extent, the same dune-to-dune spacing and the same dune heights. Yet the Martian atmospheric pressure is 1/200 that on Earth, the winds necessary to initiate the saltation of sand grains are ten times that for Earth, and the particle-size distribution may be different on the two planets. How, then, can the dune fields produced by windblown sand be so similar? What are the sources of the decameter radio emission on Jupiter, each less than 100 kilometers across, fixed on the Jovian surface, which intermittently radiate to space?

Mariner 9 observations imply that the winds on Mars at least occasionally exceed half the local speed of sound. Are the winds ever much larger? What is the nature of a transonic meteorology? There are pyramids on Mars about 3 kilometers across at the base and 1 kilometer high. They are unlikely to have been constructed by Martian pharaohs. The rate of sandblasting by wind-transported grains on Mars is at least 10,000 times that on Earth because of the greater speeds necessary to move particles in the thinner Martian atmosphere. Could the facets of the Martian pyramids have been eroded by millions of years of such sandblasting from more than one prevailing wind direction?

The moons in the outer solar system are almost certainly not replicas of our own, rather dull satellite. Many of them have such low densities that they must be composed largely of methane, ammonia or water ices. What will their surfaces look like close up? How will impact craters erode on an icy surface? Might there be volcanoes of solid ammonia with a lava of liquid NH_3 trickling down the sides? Why is Io, the innermost large satellite of Jupiter, enveloped in a cloud of gaseous sodium? How does Io help to modulate the synchrotron emission from the Jovian radiation belt in which it lives? Why is one side of Iapetus, a moon of Saturn, six times brighter than the other? Because of a particle-size difference? A chemical difference? How did such differences become established? Why on Iapetus and nowhere else in the solar system in so symmetrical a way?

The gravity of the solar system's largest moon, Titan, is so low and the temperature of its upper atmosphere sufficiently high that hydrogen should escape into space extremely rapidly in a process known as blow-off. But the spectroscopic evidence suggests that there is a substantial quantity of hydrogen on Titan. The atmosphere of Titan is a mystery. And if we go beyond the Saturnian system, we approach a region in the solar system about which we know almost nothing. Our feeble telescopes have not even reliably determined the periods of rotation of Uranus, Neptune and Pluto, much less the character of their clouds and atmospheres, and the nature of their satellite systems. The poet Diane Ackerman of Cornell University writes: "Neptune/is/elusive as a dappled horse in fog. Pulpy?/Belted? Vapory? Frost-bitten? What we know/wouldn't/fill/a lemur's fist."

One of the most tantalizing issues that we are just beginning to approach seriously is the question of organic chemistry and biology elsewhere in the solar system. The Martian environment is by no means so hostile as to exclude life, nor do we know enough about the origin and evolution of life to guarantee its presence there or anywhere else. The question of organisms both large and small on Mars is entirely open, even after the Viking missions.

The hydrogen-rich atmospheres of places such as Jupiter, Saturn, Uranus and Titan are in significant respects similar to the atmosphere of the early Earth at the time of the origin of life. From laboratory simulation experiments we know that organic molecules are produced in high yield under such conditions. In the atmospheres of Jupiter and Saturn the molecules will be convected to pyrolytic depths. But even there the steady-state concentration of organic molecules can be significant. In all simulation experiments the application of energy to such atmospheres produces a brownish polymeric material, which in many significant respects resembles the brownish coloring material in their clouds. Titan may be completely covered with a brownish, organic material. It is possible that the next few years will witness major and unexpected discoveries in the infant science of exobiology.

The principal means for the continued exploration of the solar system over the next decade or two will surely be unmanned planetary missions. Scientific space vehicles have now been launched successfully to all the planets known to the ancients. There is a range of unapproved proposed missions that have been studied in some detail. (See Chapter 16.) If most of these missions are actually implemented, it is clear that the present age of planetary exploration will continue brilliantly. But it is by no means clear that these splendid voyages of discovery will be continued, at least by the United States. Only one major planetary mission, the Galileo project to Jupiter, has been approved in the last seven years—and even it is in jeopardy.

Even a preliminary reconnaissance of the entire solar system out to Pluto and a more detailed exploration of a few planets by, for example, Mars rovers and Jupiter entry probes will not solve the fundamental problem of solar system origins; what we need is the discovery of other solar systems. Advances in ground-based and spaceborne techniques in the next two decades might be capable of detecting dozens of planetary systems orbiting nearby single stars. Recent observational studies of multiple-star systems by Helmut Abt and Saul Levy, both of Kitt Peak National Observatory, suggest that as many as one-third of the stars in the sky may have planetary companions. We do not know whether such other planetary systems will be like ours or built on very different principles.

We have entered, almost without noticing, an age of exploration and discovery unparalleled since the Renaissance. It seems to me that the practical benefits of comparative planetology for Earthbound sciences; the sense of adventure imparted by the exploration of other worlds to a society that has almost lost the opportunity for adventure; the philosophical implications of the search for a cosmic perspective—these are what will in the long run mark our time. Centuries hence, when our very real political and social problems may be as remote as the very real problems of the War of the Austrian Succession seem to us, our time may be remembered chiefly for one fact: this was the age when the inhabitants of the Earth first made contact with the cosmos around them.

CHAPTER 11

A PLANET NAMED GEORGE

◆◆◆

And teach me how
To name the bigger light, and how the less,
That burn by day and night . . .

WILLIAM SHAKESPEARE,
The Tempest, Act I, Scene 2

"Of course they answer to their names?"
the Gnat remarked carelessly.
"I never knew them to do it," [said Alice.]
"What's the use of their having names,"
said the Gnat, "if they won't answer to them?"

LEWIS CARROLL,
Through the Looking Glass

THERE IS ON the Moon a small impact crater called Galilei. It is about 9 miles across, roughly the size of the Elizabeth, New Jersey, greater metropolitan area, and is so small that a fair-sized telescope is required to see it at all. Near the center of that side of the Moon which is perpetually turned toward the Earth is a splendid ancient battered ruin of a crater, 115 miles across, called Ptolemaeus; it is easily seen with an inexpensive set of field glasses and can even be made out, by persons of keen eyesight, with the naked eye.

Ptolemy (second century A.D.) was the principal advocate of the view that our planet is immovable and at the center of the universe; he imagined that the Sun and the planets circled the Earth once daily, imbedded in swift crystalline spheres. Galileo (1564–1642), on the other hand, was a leading supporter of the Copernican view that it is the Sun which is at the center of the solar system and that the Earth is one of many planets revolving around it. Moreover, it was Galileo who, by observing the crescent phase of Venus, provided the first convincing observational evidence in favor of the Copernican view. It was Galileo who first called attention to the existence of craters on our natural satellite. Why, then, is crater Ptolemaeus so much more prominent on the Moon than crater Galileo?

The convention of naming lunar craters was established by Johannes Höwelcke, known by his Latinized name of Hevelius. A brewer and town politician in Danzig, Hevelius devoted a great deal of time to lunar cartography, publishing a famous book, *Selenographia,* in 1647. Having hand-etched the copper plates used for printing his maps of the telescopic appearance of the Moon, Hevelius was faced with the question of what to name the features depicted. Some proposed naming them after Biblical personages; others advocated philosophers and scientists. Hevelius felt that there was no logical connection between the features on the Moon and the patriarchs and prophets of thousands of years earlier, and he was also concerned that there might be substantial controversy about which philosophers and scientists—particularly if they were still alive—to honor. Taking a more prudent course, he named the prominent lunar mountains and valleys after comparable terrestrial features: as a result we have lunar Apennines, Pyrenees, Caucasus, Juras and Atlas mountains and even an Alpine valley. These names are still in use.

Galileo's impression was that the dark, flat areas on the moon were seas, real watery oceans, and that the bright and rougher regions densely studded with craters were continents. These maria (Latin for "seas") were named primarily after states of mind or conditions of nature: Mare Frigoris (the Sea of Cold), Lacus Somniorum (the Lake of Dreams), Mare Crisium (the Sea of Crises), Sinus Iridum (the Bay of Rainbows), Mare Serenitatis (the Sea of Serenity), Oceanus Procellarum (the Ocean of Storms), Mare Nubium (the Sea of Clouds), Mare Fecunditatis (the Sea of Fertility), Sinus Aestuum (the Bay of Billows), Mare Imbrium (the Sea of Rains) and Mare Tranquillitatis (the Sea of Tranquillity)—a poetic and evocative collection of place names, particularly for so inhospitable an environment as the Moon. Unfortunately, the lunar maria are bone-dry, and samples returned from them by the U.S. Apollo and Soviet Luna missions imply that never in their past were they filled with water. There never were seas, bays, lakes or rainbows on the Moon. These names have survived to the present. The first spacecraft to return data from the surface of the Moon, Luna 2, touched down in Mare Imbrium; and the first human beings to make landfall on our natural satellite, the astronauts of Apollo 11, did so, ten years later, in Mare Tranquillitatis. I think Galileo would have been surprised and pleased.

Despite Hevelius' misgivings, the lunar craters were named after scientists and philosophers by Giovanni Battista Riccioli in a 1651 publication, *Almagestum Novum*. The title of the book means "The New Almagest," the old Almagest having been the life's work of Ptolemy. ("Almagest," a modest title, means "The Greatest" in Arabic.) Riccioli simply published a map on which he placed his personal preferences for crater names, and the precedent and many of his choices have been followed without question ever since. Riccioli's book came out nine years after the death of Galileo, and there has certainly been adequate opportunity to rename craters later. Nevertheless, astronomers have retained this embarrassingly ungenerous recognition of Galileo. Twice as large as crater Galileo is one called Hell after the Jesuit father Maximilian Hell.

One of the most striking of the lunar craters is Clavius, 142 miles in diameter and the site of a fictional lunar base in the movie *2001: A Space Odyssey*. Clavius is the Latinized name of Christoffel Schlüssel (= "key" in German = Clavius), another member of the Jesuit order, and a supporter of Ptolemy. Galileo engaged in a protracted controversy on the priority of discovery and the nature of sunspots with yet another Jesuit priest, Christopher Scheiner, which developed into a bitter personal antagonism and which is thought by many historians of science to have contributed to the house arrest of Galileo, the proscription of his books, and his confession, extracted under threat of torture by the Inquisition, that his previous Copernican writings were heretical and that Earth did not move. Scheiner is commemorated by a lunar crater 70 miles across. And Hevelius, who objected altogether to the naming of lunar features after people, has a handsome crater named after himself.

Riccioli gave the names Tycho, Kepler and, interestingly, Copernicus to three of the most prominent craters on the Moon. Riccioli himself and his student Grimaldi received large craters at the limb, or edge, of the moon, Riccioli's being 106 miles across. Another prominent crater is named Alphonsus after Alphonso X of Castile, a thirteenth-century Spanish monarch who had commented, after witnessing the complexity of the Ptolemaic system, that had he been present at the Creation, he could have given God some useful suggestions on ordering the universe. (It is amusing to imagine Alphonso X's response were he to learn that seven hundred years later a nation across the Western ocean would send an engine called Ranger 9 to the Moon, automatically producing images of the lunar surface as it descended, until finally it crashed in a pre-existing depression named, after His Castilian Majesty, Alphonsus.) A somewhat less prominent crater is named after Fabricius, the Latinized name of David Goldschmidt, who in 1596 discovered that the star Mira varied periodically in brightness, striking another blow against the view championed by Aristotle and supported by the Church that the heavens were unchanging.

Thus the prejudice against Galileo in seventeenth-century Italy did not, in the naming of lunar features, carry over as a completely consistent bias in favor of Church fathers and Church doctrines on matters astronomical. Of the approximately seven thousand designated lunar formations it is difficult to extract any consistent pattern. There are craters named after political figures who had little direct or apparent connection with astronomy, such as Julius Caesar and Kaiser Wilhelm I, and after individuals of heroic obscurity: for example, crater Wurzelbaur (50 miles in diameter) and crater Billy (31 miles in diameter). Most of the designations of small lunar craters are derived from large and nearby craters, as, for example, near the crater Mösting are the smaller craters Mösting A, Mösting B, Mösting C, and so on. A wise prohibition against naming craters after living individuals has been breached only occasionally, as in assigning a few quite small craters to American astronauts of the Apollo lunar missions, and by a curious symmetry in the age of détente, to Soviet cosmonauts who remained behind in Earth orbit.

In this century an attempt has been made to name, consistently and coherently, surface features and other celestial objects by giving this function to special commissions of the International Astronomical Union (IAU), the organization of all professional astronomers on the planet Earth. A previously unnamed bay of one of the lunar "seas," examined in detail by the American Ranger spacecraft, was officially designated Mare Cognitum (the Known Sea). It is a name not so much of quiet satisfaction as of jubilation. IAU deliberations have not always been easy. For example, when the first—somewhat indistinct—photographs of the far side of the Moon were returned by the historically important Luna 3 mission, the Soviet discoverers wished to name a long, bright marking on their photographs "The Soviet Mountains." Since there is no major terrestrial mountain range of this name, the suggestion was in conflict with the Hevelius convention. It was accepted, nevertheless, in homage to the remarkable feat of Luna 3. Unfortunately, subsequent data suggest that the Soviet Mountains are not mountains at all.

In a related instance, Soviet delegates proposed naming one of the two maria on the lunar far side (both very small compared with those on the near side) Mare Moscoviense (the Sea of Moscow). But Western astronomers objected that this again departed from tradition because Moscow was neither a condition of nature nor a state of mind. It was pointed out in response that the most recent namings of lunar maria—those on the limbs, which are difficult to make out with ground-based telescopes—have not quite followed this convention either: as Mare Marginis (the Marginal Sea), Mare Orientale (the Eastern Sea) and Mare Smythii (the Smyth Sea). Perfect consistency having already been breached, the issue was decided in favor of the Soviet proposal. At an IAU meeting in Berkeley, California, in 1961, it was officially ruled by Audouin Dollfus of France that Moscow is a state of mind.

The advent of space exploration has now multiplied manyfold the problems of solar system nomenclature. An interesting example of the emerging trend can be found in the naming of features on Mars. Bright and dark surface markings on the Red Planet have been viewed, recorded and mapped from Earth for several centuries. While the nature of the markings was unknown there was an irresistible temptation to name them nevertheless. Following several abortive attempts to name them after astronomers who had studied Mars, G. V. Schiaparelli in Italy and E. M. Antoniadi, a Greek astronomer who worked in France, established around the turn of the twentieth century the convention of naming Martian features after allusions to classical mythological personages and place names. Thus we have Thoth-Nepenthes, Memnonia, Hesperia, Mare Boreum (the Northern Sea) and Mare Acidalium (the Sour Sea), as well as Utopia, Elysium, Atlantis, Lemuria, Eos (Dawn) and Uchronia (which, I suppose, can be translated as Good Times). In 1890, scholarly people were much more comfortable with classical myth than they are today.

THE KALEIDOSCOPIC surface of Mars was first revealed by American spacecraft of the Mariner series, but chiefly by Mariner 9, which orbited Mars for a full year, beginning in November 1971, and radioed back to Earth more than 7,200 close-up photographs of its surface. A profusion of unexpected and exotic detail was uncovered, including towering volcanic mountains, craters of the lunar sort but much more heavily eroded, and enigmatic, sinuous valleys which were probably caused by running water at previous epochs in the history of the planet. These new features cried out for names, and the IAU dutifully appointed a committee under the chairmanship of Gerard de Vaucouleurs of the University of Texas to propose a new Martian nomenclature. Through the efforts of several of us on the Martian nomenclature committee, a serious attempt was made to deprovincialize the new names. It was impossible to prevent major craters being named after astronomers who had studied Mars, but the range of occupations and nationalities could be significantly broadened. Thus there are Martian craters larger than 60 miles across named after the Chinese astronomers Li Fan and Liu Hsin; after biologists such as Alfred Russel Wallace, Wolf Vishniac, S. N. Vinogradsky, L. Spallanzani, F. Redi, Louis Pasteur, H. J. Muller, T. H. Huxley, J. B. S. Haldane and Charles Darwin; after a handful of geologists such as Louis Agassiz, Alfred Wegener, Charles Lyell, James Hutton and E. Suess; and even after a few science-fiction writers such as Edgar Rice Burroughs, H. G. Wells, Stanley Weinbaum and John W. Campbell, Jr. There are also two large craters on Mars named Schiaparelli and Antoniadi.

But there are many more cultures on the planet Earth—even ones with identifiable astronomical traditions—than are represented by any such list of individual names. In an attempt to offset at least in part this implicit cultural bias, a suggestion of mine was accepted to call the sinuous valleys after the names of Mars in other, largely non-European languages. On page 195 is a table of the most prominent. By a curious coincidence Ma'adim (Hebrew) and Al Qahira (Arabic: the war god after whom Cairo is named) are cheek by jowl. The landing site for the first Viking spacecraft was in Chryse, near the confluence of the Ares, Tiu, Simud and Shalbatana valleys.

TABLE 1
THE FIRST MARTIAN CHANNELS TO BE NAMED

Name	*Language*
Al Qahira	Egyptian Arabic
Ares	Greek
Auqakuh	Quechua (Inca)
Huo Hsing	Chinese
Ma'adim	Hebrew
Mangala	Sanskrit
Nirgal	Babylonian
Kasei	Japanese
Shalbatana	Akkadian
Simud	Sumerian
Tiu	Old English

For the massive Martian volcanoes, one suggestion was to name them after major terrestrial volcanoes, such as Ngorongoro or Krakatoa, which would permit some appearance on Mars of cultures with no written astronomical tradition. But this was objected to on the ground that there would be confusion when comparing terrestrial and Martian volcanoes: Which Ngorongoro are we talking about? The same potential problem exists for terrestrial cities, but we seem able to compare Portland, Oregon, with Portland, Maine, without becoming hopelessly confused. Another suggestion, made by a European savant, was to name each volcano "Mons" (mountain) followed by the name of a principal Roman deity in the appropriate Latin genitive case: thus, Mons Martes, Mons Jovis and Mons Veneris. I objected that at least the last of these had been pre-empted by quite a different field of human activity. The reply was: "Oh, I hadn't heard." The outcome was to name the Martian volcanoes after adjacent bright and dark markings in the classical nomenclature. We have Pavonis Mons, Elysium Mons and—satisfyingly, for the largest volcano in the solar system—Olympus

Mons. Thus, while the volcano names are very much in the Western tradition, by and large the most recent Mars nomenclature represents a significant break with tradition: an important number of features have been named neither after evocations of classical times nor after European geographical features and nineteenth-century Western visual astronomers.

Some Martian and lunar craters are named after the same individuals. This is the Portland case again, and I think it will cause very little confusion in practice. It does have at least one salutary benefit: on Mars there is today a large crater named Galileo. It is about the same size as the one named Ptolemaeus. And there are no craters on Mars named Scheiner or Riccioli.

Another unexpected consequence of the Mariner 9 mission is that the first close-up photographs of the moons of another planet were obtained. Maps now exist which show about half the surface features on the two Martian moons, Phobos and Deimos (the attendants of the war god, Mars). A subcommittee on Mars satellite nomenclature which I chaired assigned craters on Phobos to astronomers who had studied the moons. A prominent crater at Phobos' south pole is named after Asaph Hall, the discoverer of both moons. Astronomical apocrypha has it that Hall was on the verge of giving up his search for the Martian moons when he was directed by his wife to return to the telescope. He promptly discovered them and named them "fear" (Phobos) and "terror" (Deimos). Accordingly, the largest crater on Phobos was given Mrs. Hall's maiden name, Angelina Stickney. Had the impacting object that excavated crater Stickney been any larger, it probably would have shattered Phobos.

Deimos is reserved for writers and others who were in some way involved with speculations about the moons of Mars. The two most prominent features are named after Jonathan Swift and Voltaire, who, in their speculative romances, *Gulliver's Travels* and *Micromégas,* respectively, prefigured before the actual discovery the existence of two moons around Mars. I wanted to name a third Deimonic crater after René Magritte, the Belgian surrealist whose paintings "Le Château des Pyrénées" and "Le Sens de Réalité" pictured large rocks, suspended in the sky, of an aspect astonishingly like the two Martian moons—except for the presence in the first painting of a castle, which, so far as we know, does not surmount Phobos. The suggestion was, however, voted down as frivolous.

THIS IS THE moment in history when the features on the planets will be named forever. A crater name represents a substantial memorial: the estimated lifetime of large lunar, Martian and Mercurian craters is measured in billions of years. Because of the enormous recent increase in the number of surface features that need to be named—and also because the names of almost all dead astronomers have already been given to one or another celestial object—a new approach is needed. At the IAU meeting in Sydney, Australia, in 1973, several committees were appointed to look into questions of planetary nomenclature. One clear problem is that if craters on other planets are now named after a category other than people, we will be left with only the names of astronomers and a few others on the Moon and planets. It would be charming to name craters on, say, Mercury, after birds or butterflies, or cities or ancient vehicles of exploration and discovery. But if we accept this course, we will leave the impression on globes and maps and textbooks that we esteem only astronomers and physicists; that we care nothing for poets, composers, painters, historians, archaeologists, playwrights, mathematicians, anthropologists, sculptors, physicians, psychologists, novelists, molecular biologists, engineers and linguists. The proposal that such individuals be commemorated with unassigned lunar craters would result, say, in Dostoevsky or Mozart or Hiroshige assigned craters a tenth of a mile across, while Pitiscus is 52 miles in diameter. I do not think this would speak well for the breadth of vision and intellectual ecumenicism of the name-givers.

After a protracted debate this point of view has prevailed—in significant part due to its vigorous support by Soviet astronomers. Accordingly, the Mercury nomenclature committee, under the chairmanship of David Morrison of the University of Hawaii, has decided to name Mercurian impact craters after composers, poets and authors. Thus, major craters are named Johann Sebastian Bach, Homer and Murasaki. It is difficult for a committee of largely Western astronomers to select a group of names representative of all of world culture, and Morrison's committee requested help from appropriate musicians and experts in comparative literature. The most vexing problem is to find, for example, the names of those who composed Han dynasty music, cast Benin bronzes, carved Kwakiutl totem poles and compiled Melanesian folk epics. But even if such information comes in slowly, there will be time: the Mariner 10 photography of Mercury, which discovered the features to be named, covered only half the surface of the planet, and it will be many years before the craters in the other hemisphere will be photographed and named.

In addition, there are a few objects on Mercury that have been recommended for other sorts of names for special purposes. The proposed 20° meridian of longitude passes through a small crater which the Mariner 10 television experimenters have suggested calling Hun Kal, the Aztec word for "twenty," the base of Aztec arithmetic. And they have suggested calling an enormous depression, in some senses comparable to a lunar mare, the Caloris basin: Mercury is very hot. Finally, all of these names apply only to the topographic features of Mercury; the bright and dark markings, glimpsed dimly by past generations of ground-based astronomers, have not yet been mapped reliably. When they are, there will probably be new suggestions for naming them. Antoniadi proposed names for such features on Mercury, some of which—such as Solitudo Hermae Trismegisti (the solitude of Hermes, the thrice-great)—have a fine ring and perhaps will ultimately be retained.

NO PHOTOGRAPHIC maps of the surface of Venus exist, because the planet is perpetually enshrouded by opaque clouds. Nevertheless, surface features are being mapped by ground-based radar. Already it is apparent that there are craters and mountains, and other topographical features of stranger aspect. The success of the Venera 9 and 10 spacecraft in obtaining photographs of the planet's surface suggests that someday photographs may be returned from aircraft or balloons in the lower Venus atmosphere.

The first prominent features discovered on Venus, regions highly reflective to radar, were given unassuming names such as Alpha, Beta and Gamma. The present Venus nomenclature committee, under the chairmanship of Gordon Pettengill of the Massachusetts Institute of Technology, proposes two categories of names for Venus surface features. One category would be pioneers in radio technology whose work led to the development of the radar techniques that permit mapping the surface of Venus: for example, Faraday, Maxwell, Heinrich Hertz, Benjamin Franklin and Marconi. The other category, suggested by the name of the planet itself, would be women. At first glance, the idea of a planet devoted to women may appear sexist. But I think the opposite is true. For historical reasons, women have been discouraged from pursuing the sorts of occupations now being memorialized on other planets. The number of women after whom craters have so far been named is very small: Sklodowska (Madame Curie's maiden name); Stickney; the astronomer Maria Mitchell; the pioneer nuclear physicist Lisa Meitner; Lady Murasaki; and only a few others. While by the occupational rules for other planets women's names will continue to appear occasionally on other planetary surfaces, the Venus proposal is the only one that permits adequate recognition to be made of the historical contribution of women. (I am glad, however, that this idea will not be applied consistently; I would not myself want to see Mercury covered with businessmen and Mars with generals.)

In a fashion, women have traditionally been commemorated in the asteroid belt (see Chapter 15), that collection of rocky and metallic boulders which circle the Sun between the orbits of Mars and Jupiter. With the exception of a category of asteroids named after heroes of the Trojan War, it used to be that all asteroids were named after women. First it was largely women of classical mythology, such as Ceres, Urania, Circe and Pandora. As available goddesses dwindled, the scope broadened to include Sappho, Dike, Virginia and Sylvia. Then, as the floodgates of discovery opened and the names of astronomers' wives, mothers, sisters, mistresses and great-aunts were exhausted, they took to naming asteroids after real or hoped-for patrons and others, with a female ending appended, as, for example, Rockefelleria. By now more than two thousand asteroids have been discovered, and the situation has become moderately desperate. But non-Western traditions have hardly been tapped, and there are a multitude of Basque, Amharic, Ainu, Dobu and !Kung feminine names for future asteroids. In anticipation of an Egyptian-Israeli détente, Eleanor Helin of the California Institute of Technology proposed calling an asteroid she discovered Ra-Shalom. An additional problem—or opportunity, depending on how one views it—is that we may soon obtain close-up photographs of asteroids, with surface details that will cry out to be named.

Beyond the asteroid belt, on the planets and large moons of the outer solar system, no nondescriptive names have so far been bestowed. Jupiter, for example, has a Great Red Spot and a North Equatorial Belt, but no feature called, say, Smedley. The reason is that when we see Jupiter we are looking at its clouds, and it would not be a very fitting or at least not a very long-lived memorial to Smedley to name a cloud after him. Instead, the present major question on nomenclature in the outer solar system is what to name the moons of Jupiter. The moons of Saturn, Uranus and Neptune have satisfying or at least obscure classical names (see Table 2). But the situation for the fourteen moons of Jupiter is different.

TABLE 2
NAMES OF THE SATELLITES
OF THE OUTER PLANETS

Saturn	*Neptune*
Janus	Triton
Mimas	Nereid
Enceladus	
Tethys	*Uranus*
Dione	Miranda
Rhea	Ariel
Titan	Umbriel
Hyperion	Titania
Iapetus	Oberon
Phoebe	
	Pluto
	Charon

The four large moons of Jupiter were discovered by Galileo, whose theological contemporaries were convinced by a vague amalgam of Aristotelian and Biblical ideas that the other planets could have no moons. The contrary discovery by Galileo was disconcerting to fundamentalist churchmen of the time. Possibly in an effort to circumvent criticism, Galileo called the moons the Medicean satellites—after his funding agency. But posterity has been wiser: they are known instead as the Galilean satellites. In a similar vein, when William Herschel of England discovered the seventh planet he proposed calling it George. If wiser heads had not prevailed, we might today have a major planet named after George III. Instead we call it Uranus.

The Galilean satellites were assigned their Greek mythological names by Simon Marius (commemorated on the Moon by a crater 27 miles across), a contemporary of Galileo and a disputant with him for the priority of their discovery. Marius and Johannes Kepler felt that it would be extremely unwise to name celestial objects after real people and particularly after political personages. Marius wrote: "I want the thing done without superstition and with the sanction of theologians. Jupiter especially is charged by the poets with illicit loves. Especially well-known among these are three virgins, whose love Jupiter secretly coveted and obtained, namely: Io . . . Callisto . . . and Europa . . . Yet even more ardently did he love the beautiful boy Ganymede . . . and so I believe that I have not done badly in naming the first Io, the second Europa, the third, on account of the splendor of its light, Ganymede, and lastly the fourth Callisto."

However, in 1892 E. E. Barnard discovered a fifth moon of Jupiter with an orbit interior to Io's. Barnard resolutely insisted that this satellite should be called Jupiter 5 and by no other name. Since then, Barnard's position has been maintained, and of the fourteen Jovian moons now known, only the Galilean satellites had, until recently, names officially sanctioned by the IAU. However unreasonable it may be, people show a strong preference for names over numbers. (This is clearly illustrated in the resistance of college students to being considered "only a number" by the college bursar; by the outrage of many citizens at being known to the government only by their social security number; and by the systematic attempts in jails and prison camps to demoralize and degrade the inmates by assigning them a numeral as their only identity.) Soon after Barnard's discovery, Camille Flammarion suggested the name Amalthea for Jupiter 5 (Amalthea was in Greek legend the goat that suckled the infant Zeus). While being suckled by a goat is not precisely an act of illicit love, it must have seemed, to the Gallic astronomer, adequately close.

The IAU committee on Jovian nomenclature, chaired by Tobias Owen of the State University of New York at Stony Brook, has proposed a set of names for Jupiter 6 through 13. Two principles guided their selection: the name chosen should be that of "an illicit love" of Jupiter, but one so obscure as to have been missed by those indefatigable cullers of the classics who name asteroids, and must end with an *a* or an *e* depending on whether the moon goes around Jupiter clockwise or counterclockwise. But in the opinion of at least some classical scholars, these names are obscure to the point of bewilderment, and the result leaves many of the most prominent Jovian paramours unrepresented in the Jupiter system. The result is particularly poignant in that Hera (Juno), the wife so often scorned by Zeus (Jupiter), is not represented at all. Evidently, she was inadequately illicit. An alternative list of names, which includes most of the prominent paramours as well as Hera, is also shown in the table below. Were these names employed, it is true they would duplicate asteroid names. This is in any case already a fact for the four Galilean satellites, where the amount of confusion thus engendered has been negligible. On the other hand, there are those who support Barnard's position that numbers are sufficient; prominent among these is Charles Kowal* of the California Institute of Technology, the discoverer of Jupiter 13 and Jupiter 14. There seems to be merit in all three positions and it will be interesting to see how the debate turns out. At least we do not yet have to judge the merits of contending suggestions for naming features on the Jovian satellites.

TABLE 3

PROPOSED NAMES FOR JOVIAN SATELLITES

Satellite	*I.A.U. Committee Names*	*Alternative Names Suggested Here*
J V	Amalthea	Amalthea
VI	Himalia	Maia
VII	Elara	Hera
VIII	Pasiphaë	Alcmene
IX	Sinope	Leto
X	Lysithea	Demeter
XI	Carme	Semele
XII	Anake	Danaë
XIII	Leda	Leda
XIV	—	—

* Kowal has also recently discovered a very interesting small object orbiting the Sun between the orbits of Uranus and Saturn. It may be the largest member of a new asteroid belt. Kowal proposes calling it Chiron, after the centaur who educated many Greek mythological gods and heroes. If other trans-Saturnian asteroids are discovered, they can be named after other centaurs.

But that time is not long off. There are thirty-one known moons of Jupiter, Saturn, Uranus and Neptune. None has been photographed close up. The decision has recently been made to name features on the moons in the outer solar system after mythological figures from all cultures. However, very soon the Voyager mission will obtain high-resolution images of about ten of them, in addition to the rings of Saturn. The total surface area of the small objects in the outer solar system greatly exceeds the areas of Mercury, Venus, Earth, Moon, Mars, Phobos and Deimos together. There will be ample opportunity for all human occupations and cultures to be represented eventually, and I daresay provisions for nonhuman species can also be made. There are probably more professional astronomers alive today than in the total prior recorded history of mankind. I suppose that many of us will also be commemorated in the outer solar system—a crater on Callisto, a volcano on Titan, a ridge on Miranda, a glacier on Halley's comet. (Comets, incidentally, are given the names of their discoverers.) I sometimes wonder what the arrangement will be—whether those who are bitter rivals will be separated by being placed on different worlds, and whether those whose discoveries were collaborative will nestle together, crater rampart to crater rampart. There have been objections that political philosophers are too controversial. I myself would be delighted to see two enormous, adjacent craters called Adam Smith and Karl Marx. There are even enough objects in the solar system for dead political and military leaders to be accommodated. There are those who have advocated supporting astronomy by selling crater names to the highest bidders, but I think this goes rather too far.

THERE IS A curious problem about names in the outer solar system. Many of the objects there have extremely low density, as if they were made of ice, great fluffy snowballs tens or hundreds of miles across. While objects impacting these bodies will certainly produce craters, craters in ice will not last very long. At least for some objects in the outer solar system, named features may be transient. Perhaps that is a good thing: it would give us a chance to revise our opinions of politicians and others, and will give eventual recourse if flushes of national or ideological fervor are reflected in solar system nomenclature. The history of astronomy shows that some suggestions for celestial nomenclature are better ignored. For example, in 1688 Erhard Weigel at Jena proposed a revision of the ordinary zodiacal constellations—the lion, virgin, fish and water carrier that people have in mind when they ask you what "sign" you are. Weigel proposed instead a "heraldic sky" in which the royal families of Europe would be represented by their tutelary animals: a lion and a unicorn for England, for example. I hate to imagine descriptive stellar astronomy today had that idea been adopted in the seventeenth century. The sky would be carved into two hundred tiny patches, one for each nation-state existing at the time.

The naming of the solar system is fundamentally not a task for the exact sciences. It has historically encountered prejudice and jingoism and lack of foresight at every turn. However, while it may be a little early for self-congratulation, I think astronomers have recently taken some major steps to deprovincialize the nomenclature and make it representative of all of humanity. There are those who think it is a pointless, or at least thankless, task. But some of us are convinced it is important. Our remote descendants will be using our nomenclature for their homes: on the broiling surface of Mercury; by the banks of the Martian valleys; on the slopes of Titanian volcanoes; or on the frozen landscape of distant Pluto, where the Sun appears as a point of bright light in a sky of unremitting blackness. Their view of us, of what we cherish and hold dear, may be determined largely by how we name the moons and planets today.

CHAPTER 12

LIFE IN THE SOLAR SYSTEM

◆◆◆

"I see nobody on the road," said Alice.

"I only wish *I* had such eyes," the King remarked in a fretful tone. "To be able to see Nobody! And at that distance too! Why, it's as much as *I* can do to see real people, by this light!"

LEWIS CARROLL,
Through the Looking Glass

MORE THAN three hundred years ago, Anton van Leeuwenhoek of Delft explored a new world. With the first microscope he viewed a stagnant infusion of hay and was astounded to find it swarming with small creatures:

On April 24th, 1676, observing this water by chance, I saw therein with great wonder unbelievably very many small animalcules of various sorts; among others, some that were three to four times as long as broad. Their entire thickness was, in my judgement, not much thicker than one of the little hairs that cover the body of a louse. These creatures had very short, thin legs in front of the head (although I can recognize no head, I speak of the head for the reason that this part always went forward during movement) . . . Close to the hindmost part lay a clear globule; and I judged that the very hindmost part was slightly cleft. These animalcules are very cute while moving about, oftentimes tumbling all over.

These tiny "animalcules" had never before been seen by any human being. Yet Leeuwenhoek had no difficulty in recognizing them as alive.

Two centuries later Louis Pasteur developed the germ theory of disease from Leeuwenhoek's discovery and laid the foundation for much of modern medicine. Leeuwenhoek's objectives were not practical at all, but exploratory and adventuresome. He himself never guessed the future practical applications of his work.

In May of 1974 the Royal Society of Great Britain held a discussion meeting on "The Recognition of Alien Life." Life on Earth has developed by a slow, tortuous step-by-step progression known as evolution by natural selection. Random factors play a critical role in this process—as, for example, which gene at what time will be mutated or changed by an ultraviolet photon or a cosmic ray from space. All the organisms on Earth are exquisitely adapted to the vagaries of their natural environments. On some other planet, with different random factors operating and extremely exotic environments, life may have evolved very differently. If we landed a spacecraft on the planet Mars, for example, would we even be able to recognize the local life forms as alive?

One theme which was stressed at the Royal Society discussion was that life elsewhere should be recognizable by its improbability. Take trees, for example. Trees are long skinny structures, above ground fatter at the top than at the bottom. It is easy to see that after millennia of rubbing by wind and water, most trees should have fallen down. They are in mechanical disequilibrium. They are unlikely structures. Not all top-heavy structures are produced by biology. There are, for example, pedestal rocks in deserts. But were we to see a great many top-heavy structures, all closely similar, we could make a reasonable guess that they were of biological origin. Likewise for Leeuwenhoek's animalcules. There are many of them, closely similar, highly complex and improbable in the extreme. Without ever having seen them before, we correctly guess they are biological.

There have been elaborate debates on the nature and definition of life. The most successful definitions invoke the evolutionary process. But we do not land on another planet and wait to see if any nearby objects evolve. We do not have the time. The search for life then takes on a much more practical aspect. This point was brought out with some finesse at the Royal Society discussion when, after an exchange remarkable for its rambling metaphysical vagueness, Sir Peter Medawar rose to his feet and said, "Gentlemen, everyone in this room knows the difference between a live horse and a dead horse. Pray, therefore, let us cease flogging the latter." Medawar and Leeuwenhoek would have seen eye to eye.

But are there trees or animalcules on the other worlds of our solar system? The simple answer is that no one yet knows. From the vantage point of the nearest planets, it would be impossible to detect photographically the presence of life on our own planet. Even from the closest orbital observations of Mars made to date, from the American spacecraft Mariner 9 and Viking 1 and 2, details on Mars much smaller than 100 meters across have remained invisible. Since even the most ardent enthusiasts of extraterrestrial life do not anticipate Martian elephants 100 meters long, many important tests have not yet been performed.

At the present time we can only assess the physical environments of the other planets, determine whether they are so severe as to exclude life—even forms rather different from those we know on Earth—and in the case of the more clement environments perhaps speculate on the life forms that might be present. The one exception is the Viking lander results, briefly discussed below.

A place may be too hot or too cold for life. If the

temperatures are very high—say, several thousands of degrees Centigrade—then the molecules that make up the organism will fall to pieces. Thus it is customary to exclude the Sun as an abode of life. On the other hand, if the temperatures are too low, then the chemical reactions that drive the internal metabolism of the organism will proceed at too ponderous a pace. For this reason the frigid wastes of Pluto are customarily excluded as an abode of life. However, there may be chemical reactions which proceed at respectable rates at low temperatures but which are unexplored here on Earth, where chemists dislike working in laboratories at –230°C. We must be careful not to take too chauvinistic a view of the matter.

The giant outer planets of the solar system, Jupiter, Saturn, Uranus, and Neptune, are sometimes excluded from biological considerations because their temperatures are very low. But these temperatures are the temperatures of their upper clouds. Deeper down in the atmospheres of such planets, as in the atmosphere of the Earth, much more clement conditions are to be encountered. And they appear to be rich in organic molecules. By no means can they be excluded.

While we human beings enjoy oxygen, this is hardly a recommendation for it, since there are many organisms that are poisoned by it. If the thin protective ozone layer in our atmosphere, made by sunlight from oxygen, did not exist, we would rapidly be fried by ultraviolet light from the Sun. But on other worlds, ultraviolet sunshades or biological molecules impervious to near-ultraviolet radiation can readily be imagined. Such considerations merely underline our ignorance.

An important distinction among the other worlds of our solar system is the thickness of their atmospheres. In the total absence of an atmosphere it is very difficult to conceive of life. As on Earth, the biology on other planets must, we think, be driven by sunlight. On our planet, the plants eat the sunlight and the animals eat the plants. Were all the organisms on Earth forced (by some unimaginable catastrophe) into a subterranean existence, life would cease as soon as accumulated food stores were exhausted. The plants, the fundamental organisms on any planet, must see the Sun. But if a planet has no atmosphere, not only ultraviolet radiation but X-rays and gamma rays and charged particles from the solar wind will fall unimpeded on the planetary surface and frizzle the plants.

Furthermore, an atmosphere is necessary for exchange of materials so that the basic molecules for biology are not all used up. On Earth, for example, green plants give off oxygen—a waste product—into the atmosphere. Many respiring animals, like human beings, breathe the oxygen and give off carbon dioxide, which the plants in turn imbibe. Without this clever (and painfully evolved) equilibrium between plants and animals, we would rapidly run out of oxygen or carbon dioxide. For these two reasons—radiation protection and molecular exchange—an atmosphere seems required for life.

Some of the worlds in our solar system have exceedingly thin atmospheres. Our Moon, for example, has at its surface less than one million millionth the atmospheric pressure on Earth. Six places on the near side of the Moon were examined by Apollo astronauts. No top-heavy structures, no lumbering beasts were found. Nearly four hundred kilograms of samples have been returned from the Moon and meticulously examined in terrestrial laboratories. There were no animalcules, no microbes, almost no organic chemicals, or even any water. We expected the Moon to be lifeless, and apparently it is. Mercury, the closest planet to the Sun, resembles the Moon. Its atmosphere is exceedingly thin, and it ought not to support life. In the outer solar system there are many large satellites the size of Mercury or our own Moon, composed of some mix of rock (like the Moon and Mercury) and ices. Io, the second moon of Jupiter, falls into this category. Its surface seems to be covered with a kind of reddish salt deposit. We are very ignorant about it. But because of its very low atmospheric pressure, we do not expect life on it.

Then there are planets with moderate atmospheres. Earth is the most familiar example. Here life has played a major role in determining the composition of our atmosphere. The oxygen is, of course, produced by green-plant photosynthesis, but even the nitrogen is thought to be made by bacteria. Oxygen and nitrogen together comprise 99 percent of our atmosphere, which has evidently been reworked on a massive scale by the life on our planet.

The total pressure on Mars is about one half of one percent that on Earth, but the atmosphere there is composed largely of carbon dioxide. There are small quantities of oxygen, water vapor, nitrogen and other gases. The Martian atmosphere has not obviously been reworked by biology, but we do not know Mars well enough to exclude life there. It has congenial temperatures at some times and places, a dense enough atmosphere, and abundant water locked away in the ground and polar caps. Even some varieties of terrestrial microorganisms can survive there very well. Mariner 9 and Viking found hundreds of dry riverbeds, apparently indicating a time in the recent geological history of the planet when abundant liquid water flowed. It is a world awaiting exploration.

A third and less familiar example of places with moderate atmospheres is Titan, the largest moon of Saturn. Titan appears to have an atmosphere with a density between that of Mars and Earth. This atmosphere is, however, composed largely of hydrogen and methane, and is surmounted by an unbroken layer of reddish clouds—probably complex organic molecules. Because of its remoteness, Titan has attracted the interest of exobiologists only recently, but it holds the promise of a long-term fascination.

The planets with very dense atmospheres present a special problem. Like Earth, their atmospheres are cold at the top and warmer at the bottom. But when the atmosphere is very thick, the temperatures at the bottom become too hot for biology. In the case of Venus, the surface temperatures are about 480°C; for the Jovian planets, many thousands of degrees Centigrade. All these atmospheres, we think, are convective, with vertical winds vigorously carrying materials both up and down. Life probably cannot be imagined on their surfaces because of the high temperatures. The cloud environments are perfectly clement, but convection will carry hypothetical cloud organisms down to the depth and fry them there. There are two obvious solutions. There might be small organisms that reproduce as fast as they are carried down to the planetary skillet or the organisms might be buoyant. Fish on Earth have float bladders for a similar purpose, and both on Venus and on the Jovian planets, organisms that are essentially hydrogen-filled balloons can be envisioned. For them to float at modest temperatures on Venus, they need to be at least a few centimeters across, but for the same purpose on Jupiter, they must be at least meters across—the size of ping-pong balls and meteorological balloons, respectively. We do not know that such beasts exist, but it is of some little interest to see that they can be envisioned without doing violence to what is known of physics, chemistry or biology.

Our profound ignorance of whether other planets harbor life may end within this century. Plans are now afoot for the chemical and biological examination of many of these candidate worlds. The first step was the American Viking missions, which landed two sophisticated automatic laboratories on Mars in the summer of 1976, almost three hundred years to the month of Leeuwenhoek's discovery of hay infusoria. Viking found no curious structures nearby (or sauntering by) which were top-heavy, and no detectable organic molecules. Of three experiments in microbial metabolism, two in both landing sites repeatedly gave what seemed to be positive results. The implications are still under vigorous debate. In addition, we must remember that the two Viking landers examined closely, even with photography, less than one millionth of the surface area of the planet. More observations—particularly with more sophisticated instrumentation (including microscopes) and with roving vehicles—are needed. But despite the ambiguous nature of the Viking results, these missions represent the first time in the history of the human species that another world has been seriously examined for life.

In the following decades it is likely that there will be buoyant probes into the atmospheres of Venus, Jupiter and Saturn, and landers on Titan, as well as more detailed studies of the surface of Mars. A new age of planetary exploration and exobiology dawned in the seventh decade of the twentieth century. We live in a time of adventure and high intellectual excitement; but also—as the step from Leeuwenhoek to Pasteur shows—in the midst of an endeavor which promises great practical benefits.

CHAPTER 13

TITAN, THE ENIGMATIC MOON OF SATURN

◆◆◆

On Titan, warmed by a hydrogen blanket,
ice-ribbed volcanoes jet ammonia
dredged out of a glacial heart. Liquid
and frozen assets uphold an empire
bigger than Mercury, and even a little
like primitive Earth: asphalt plains and hot
mineral ponds. But
how I'd like to take the waters of Titan, under
that fume-ridden sky,
where the land's blurred by cherry mist
and high above, like floating wombs,
clouds
tower and swarm, raining down primeval
bisque, while life waits in the wings.

DIANE ACKERMAN,
The Planets (New York, Morrow, 1976)

TITAN IS NOT a household word, or world. We do not usually think of it when we run through a list of familiar objects in the solar system. But in the last few years this satellite of Saturn has emerged as a place of extraordinary interest and prime significance for future exploration. Our most recent studies of Titan have revealed that it has an atmosphere more like the Earth's—at least in terms of density—than any other object in the solar system. This fact alone gives it new significance as the exploration of other worlds begins in earnest.

Besides being the largest satellite of Saturn, Titan is also, according to recent work by Joseph Veverka, James Elliot and others at Cornell University, the largest satellite in the solar system—about 5,800 kilometers (3,600 miles) in diameter. Titan is larger than Mercury and nearly as large as Mars. And yet there it is in orbit around Saturn.

We might obtain some clues about the nature of Titan by examining the two major worlds in the outer solar system—Jupiter and Saturn. Both have a general reddish or brownish coloration. That is, the upper layer of clouds that we see from the Earth has this hue primarily. Something in the atmosphere and clouds of these planets is strongly absorbing blue and ultraviolet light, so that the light that is reflected back to us is primarily red. The outer solar system, in fact, has a number of objects that are remarkably red. Although we have no color photographs of Titan because it is 800 million miles away and has an angular size smaller than the Galilean satellites of Jupiter, photoelectric studies reveal that it is, in fact, very red. Astronomers who thought about the problem once believed that Titan was red for the same reason that Mars is red: a rusty surface. But then the reason for Titan's red color would be different from the reason for Jupiter's and Saturn's, because we do not see to a solid surface on those planets.

In 1944 Gerard Kuiper detected spectroscopically an atmosphere of methane around Titan—the first satellite found to have an atmosphere. Since then, the methane observations have been confirmed, and at least moderately suggestive evidence for the presence of molecular hydrogen has been provided by Lawrence Trafton of the University of Texas.

Since we know the amount of gas necessary to produce the observed spectral absorption features, and we know from its mass and radius the surface gravity of Titan, we can deduce the minimum atmospheric pressure. We find it is something like 10 millibars, about one percent of the Earth's atmospheric pressure—a pressure that exceeds that of Mars. Titan has the most Earth-like atmospheric pressure in the solar system.

Not only the best, but the only visual telescopic observations of Titan have been made by Audouin Dollfus at the Meudon Observatory in France. These are hand drawings done at the telescope during moments of atmospheric steadiness. From the variable patches that he observed, Dollfus concluded that things are happening on Titan that do not correlate with the satellite's rotation period. (Titan is thought always to face Saturn, as our Moon does the Earth.) Dollfus guessed that there might be clouds, at least of a patchy sort, on Titan.

Our knowledge of Titan has made a number of substantial quantum jumps forward in recent years. Astronomers have successfully obtained the polarization curve of small objects. The idea is that initially unpolarized sunlight falls on Titan, say, and is polarized on reflection. The polarization is detected by a device similar in principle to, but more sophisticated and sensitive than, "polaroid" sunglasses. The amount of polarization is measured as Titan goes through a small range of phases—between "full" Titan and slightly "gibbous" Titan. The resulting polarization curve, when compared to laboratory polarization curves, gives information on the size and composition of the material responsible for the polarization.

The first polarization observations of Titan, made by Joseph Veverka, indicated that the sunlight reflected back from Titan is most likely reflected off clouds and not off a solid surface. Apparently there is on Titan a surface and a lower atmosphere that we do not see; an opaque cloud deck and an overlying atmosphere, both of which we do see; and an occasional patchy cloud above that. Since Titan appears red, and we view it at the cloud deck, there must, according to this argument, be red clouds on Titan.

Additional support for this concept comes from the extremely low amount of ultraviolet light reflected from Titan, as measured by the Orbiting Astronomical Observatory. The only way to keep Titan's ultraviolet brightness small is to have the ultraviolet absorbing stuff high up in the atmosphere. Otherwise Rayleigh scattering by the atmospheric molecules themselves would make Titan bright in the ultraviolet. (Rayleigh scattering is the preferential scattering of blue rather than red light, which is responsible for blue skies on Earth.)

But material that absorbs in the ultraviolet and violet appears red in reflected light. So there are two separate lines of evidence (or three, if we believe the hand drawings) for an extensive cloud cover on Titan. What do we mean by extensive? More than 90 percent of Titan must be cloaked in clouds to match the polarization data. Titan seems to be covered by dense red clouds.

A second astonishing development was inaugurated in 1971 when D. A. Allen of Cambridge University and T. L. Murdock of the University of Minnesota found that the observed infrared emission from Titan at a wavelength of 10 to 14 microns is more than twice what is expected from solar heating. Titan is too small to have a significant internal energy source like Jupiter or Saturn. The only explanation seemed to be the greenhouse effect in which the surface temperature rises until the infrared radiation trickling out just balances the absorbed visible radiation coming in. It is the greenhouse effect that keeps the surface temperature of the Earth above freezing and the temperature of Venus at 480°C.

But what could cause a Titanian greenhouse effect? It is unlikely to be carbon dioxide and water vapor as on Earth and Venus, because these gases should be largely frozen out on Titan. I have calculated that a few hundred millibars of hydrogen (1,000 millibars is the total sea-level atmospheric pressure on Earth) would provide an adequate greenhouse effect. Since this is more than the amount of hydrogen observed, the clouds would have to be opaque at certain short wavelengths and more nearly transparent at certain longer wavelengths. James Pollack, at NASA's Ames Research Center, has calculated that a few hundred millibars of methane might also be adequate and, moreover, might explain some of the details of the infrared emission spectrum of Titan. This large amount of methane would also have to hide under the clouds. Both greenhouse models have the virtue of invoking only gases thought to exist on Titan; of course, both gases might play a role.

An alternative model of the Titan atmosphere was proposed by the late Robert Danielson and his colleagues at Princeton University. They suggest that small quantities of simple hydrocarbons—such as ethane, ethylene and acetylene—which have been observed in the upper atmosphere of Titan absorb ultraviolet light from the Sun and heat the upper atmosphere. It is then the hot upper atmosphere and not the surface that we see in the infrared. On this model there need be no enigmatically warm surface, no greenhouse effect, and no atmospheric pressure of hundreds of millibars.

Which view is correct? At the present time no one knows. The situation is reminiscent of studies of Venus in the early 1960s when the planet's radio-brightness temperature was known to be high, but whether the emission was from a hot surface or a hot region of the atmosphere was (appropriately) hotly debated. Since radio waves pass through all but the densest atmospheres and clouds, the Titan problem might be resolved if we had a reliable measure of the radio-brightness temperature of the satellite. The first such measurement was performed by Frank Briggs of Cornell with the giant interferometer of the National Radio Astronomy Observatory in Green Bank, West Virginia. Briggs finds a surface temperature of Titan of –140°C with an uncertainty of 45°. The temperature in the absence of a greenhouse effect is expected to be about –185°C. Briggs's observations therefore seem to suggest a fairly sizable greenhouse effect and a dense atmosphere, but the probable error of the measurements is still so large as to permit the zero greenhouse case.

Subsequent observations by two other radio astronomical groups give values both higher and lower than Briggs's results. The higher range of temperatures, astonishingly, even approaches temperatures in cold regions of the Earth. The observational situation, like the atmosphere of Titan, seems very murky. The problem could be resolved if we could measure the size of the solid surface of Titan by radar (optical measurements give us the distance from cloudtop to cloudtop). The problem may have to await studies by the Voyager mission, which is scheduled to send two sophisticated spacecraft by Titan—one very close to it—in 1981.

Whichever model we select is consistent with the red clouds. But what are they made of? If we take an atmosphere of methane and hydrogen and supply energy to it, we will make a range of organic compounds, both simple hydrocarbons (like the sort that are needed to make Danielson's inversion layer in the upper atmosphere) and complex ones. In our laboratory at Cornell, Bishun Khare and I have simulated the kinds of atmospheres that exist in the outer solar system. The complex organic molecules we synthesize in them have optical properties similar to those of the Titanian clouds. We think there is strong evidence for abundant organic compounds on Titan, both simple gases in the atmosphere and more complex organics in the clouds and on the surface.

One problem with an extensive Titanian atmosphere is that the light gas hydrogen should be gushing away because of the low gravity. The only way that I can explain this situation is that the hydrogen is in a "steady state." That is, it escapes but is replenished from some internal source—volcanoes, most likely. The density of Titan is so low that its interior must be almost entirely composed of ices. We can think of it as a giant comet made of methane, ammonia and water ices. There must also be a small admixture of radioactive elements which, while decaying, will heat their surroundings. The heat conduction problem has been worked out by John Lewis, of MIT, and it is clear that the near-surface interior of Titan will be slushy. Methane, ammonia and water vapor should be outgassed from the interior and broken down by ultraviolet sunlight, producing atmospheric hydrogen and cloud organic compounds at the same time. There may be surface volcanoes made of ice instead of rock, spewing out in occasional eruptions not liquid rock but liquid ice—a lava of running methane, ammonia and perhaps water.

There is another consequence of the escape of all this hydrogen. An atmospheric molecule that achieves escape velocity from Titan generally does not have escape velocity from Saturn. Thus, as Thomas McDonough and the late Neil Brice of Cornell have pointed out, the hydrogen that is being lost from Titan will form a diffuse toroid, or doughnut, of hydrogen gas around Saturn. This is a very interesting prediction, first made for Titan but possibly relevant for other satellites as well. Pioneer 10 has detected such a hydrogen toroid around Jupiter in the vicinity of Io. As Pioneer 11 and Voyager 1 and 2 fly near Titan, they may be able to detect the Titan toroid.

Titan will be the easiest object to explore in the outer solar system. Nearly atmosphereless worlds such as Io or the asteroids present a landing problem because we cannot use atmospheric braking. Giant worlds such as Jupiter and Saturn have the opposite problem: the acceleration due to gravity is so large and the increase in atmospheric density is so rapid that it is difficult to devise an atmospheric probe that will not burn up on entry. Titan, however, has a dense enough atmosphere and a low enough gravity. If it were a little closer, we probably would be launching entry probes there today.

Titan is a lovely, baffling and instructive world which we suddenly realize is accessible for exploration: by fly-bys to determine the gross global parameters and to search for breaks in the clouds; by entry probes to sample the red clouds and unknown atmosphere; and by landers to examine a surface like none we know. Titan provides a remarkable opportunity to study the kinds of organic chemistry that on Earth may have led to the origin of life. Despite the low temperatures, it is by no means impossible that there is a Titanian biology. The geology of the surface may be unique in all the solar system. Titan is waiting . . .

CHAPTER 14

THE CLIMATES OF PLANETS

◆◆◆

Is it not the height of silent humour
To cause an unknown change
in the earth's climate?

ROBERT GRAVES,
The Meeting

BETWEEN 30 and 10 million years ago, it is thought, temperatures on Earth slowly declined, by just a few Centigrade degrees. But many plants and animals have their life cycles sensitively attuned to the temperature, and vast forests receded toward more tropical latitudes. The retreat of the forests slowly removed the habitats of small furry binocular creatures, weighing only a few pounds, which had lived out their days brachiating from branch to branch. With the forests gone, only those furry creatures able to survive on the grassy savannas were to be found. Some tens of millions of years later, those creatures left two groups of descendants: one which includes the baboons and the other called humans. We may owe our very existence to climatic changes that on the average amount to only a few degrees. Such changes have brought some species into being and extinguished others. The character of life on our planet has been powerfully influenced by such variations, and it is becoming increasingly clear that the climate is continuing to change today.

There are many indications of past climatic changes. Some methods reach far into the past, others have only a limited applicability. The reliability of the methods also differs. One approach, which may be valid for a million years back in time, is based on the ratio of the isotopes oxygen 18 to oxygen 16 in the carbonates of shells of fossil foraminifera. These shells, belonging to species very similar to some that can be studied today, vary the oxygen 16/oxygen 18 ratio according to the temperature of the water in which they grew. Somewhat similar to the oxygen-isotope method is one based upon the ratio of the isotopes sulfur 34 to sulfur 32. There are other, more direct fossil indicators; for example, the widespread presence of corals, figs and palms denotes high temperatures, and the abundant remains of large hairy beasts, such as mammoths, indicate cold temperatures. The geological record is replete with extensive evidence of glaciation—great moving sheets of ice that leave characteristic boulders and erosional traces. There is also clear geological evidence for beds of evaporites—regions where briny water has evaporated leaving behind the salts. Such evaporation occurs preferentially in warm climates.

When this range of climatic information is put together, a complex pattern of temperature variation emerges. At no time, for example, is the average temperature of the Earth below the freezing point of water, and at no time does it even approach the normal boiling point of water. But variations of several degrees are common, and even variations of twenty or thirty degrees may have occurred at least locally. Fluctuations of a few degrees Centigrade happen over characteristic times of tens of thousands of years, and the recent succession of glacial and interglacial periods has this timing and temperature amplitude. But there are climatic fluctuations over much longer periods, the longest being on the order of a few hundred million years. Warm periods appear to have occurred about 650 million years ago and 270 million years ago. By the standards of past climatic fluctuations, we are now in the midst of an ice age. For most of the Earth's history, there were no "permanent" ice caps, as in the Arctic and Antarctic today. We have, over the past few hundred years, made a partial emergence from our ice age caused by some as yet unexplained minor climatic variation; and there are certain signs that we may plunge back into the global cold temperatures characteristic of our epoch as seen from the perspective of the immense vistas of geological time. It is a sobering fact that 2 million years ago the site of the city of Chicago was buried under a mile of ice.

What determines the temperature of Earth? As seen from space, it is a rotating blue ball streaked with varying cloud patches, reddish-brown deserts and brilliant white polar caps. The energy for heating the Earth comes almost exclusively from sunlight, the energy conducted up from the hot interior of the Earth amounting to less than one thousandth of one percent of that arriving in the form of visible light from the Sun. But not all the sunlight is absorbed by the Earth. Some is reflected back to space by polar ice, clouds, and the rocks and water on the surface of the Earth. The average reflectivity, or albedo, of the Earth, as measured directly from satellites and indirectly from Earthshine reflected off the dark side of the Moon, is about 35 percent. The 65 percent of sunlight that is absorbed by the Earth heats it to a temperature which can readily be calculated. This temperature is about –18°C, below the freezing point of seawater and some 30°C colder than the measured average temperature of the Earth.

The discrepancy is due to the fact that this calculation neglects the so-called greenhouse effect. Visible light from the Sun enters the Earth's clear atmosphere and is transmitted through to the surface. The surface, however, in attempting to radiate back into space, is constrained by the laws of physics to do so in the infrared. The atmosphere is not so transparent in the infrared, and at some wavelengths of infrared radiation—such as 6.2 microns or 15 microns—radiation would travel only a few centimeters before being absorbed by atmospheric gases. Since the Earth's atmosphere is murky and absorbing at many wavelengths in the infrared, the thermal radiation given off by the surface of the Earth is impeded in escaping to space. In order to have a close equality between the radiation received by the Earth from the Sun and the radiation emitted by the Earth to space, the surface temperature of the Earth must then rise. The greenhouse effect is due not to the major atmospheric constituents of the Earth, such as oxygen and nitrogen, but almost exclusively to the minor constituents, especially carbon dioxide and water vapor.

As we have seen, the planet Venus is probably a case where the massive injection of carbon dioxide and smaller amounts of water vapor into a planetary atmosphere has led to such a large greenhouse effect that water cannot be maintained on the surface in the liquid state; hence, the planetary temperature runs away to some extremely high value—in the case of Venus, 480°C.

We have so far been talking about average temperatures. The temperature of the Earth varies from place to place. It is colder at the poles than at the equator because, in general, sunlight falls directly on the equator and obliquely on the poles. The tendency for the temperatures to be very different between equator and poles on Earth is moderated by atmospheric circulation. Hot air rises at the equator and moves at high altitudes to the poles, where it settles and returns to the surface; it then retraces its path, but at low altitudes, from pole back to equator. This general motion—complicated by the rotation of the Earth, its topography and the phase changes of water—is responsible for weather.

The observed average temperature of about 15°C on the Earth today can be explained quite well by the observed intensity of sunlight, global albedo, the tilt of the rotational axis and the greenhouse effect. But all of these parameters can, in principle, vary; and past or future climatic change can be attributed to changes in any of them. In fact, there have been almost a hundred different theories of climatic change on Earth, and even today the subject is hardly marked by unanimity of opinion. This is not because climatologists are by nature ignorant or contentious, but rather because the subject is exceedingly complex.

Both negative and positive feedback mechanisms probably exist. Suppose, for example, there were a decrease of a few degrees in the Earth's temperature. The amount of water vapor in the atmosphere is determined almost entirely by temperature and declines by snowing out as the temperature declines. Less water in the atmosphere implies a smaller greenhouse effect and a further lowering of the temperature, which may result in even less atmospheric water vapor, and so on. Likewise, a decline in temperature may increase the amount of polar ice, increasing the albedo of the Earth and decreasing the temperature still further. On the other hand, a decline in temperature may decrease the amount of cloudiness, which will decrease the average albedo of the Earth and increase the temperature—perhaps enough to undo the initial temperature decrease. And it has been proposed recently that the biology of the planet Earth acts as a kind of thermostat to prevent too extreme excursions in temperature which might have deleterious global biological consequences. For example, a decline in temperature may cause an increase of a species of hardy plants that has extensive ground cover and low albedo.

Three of the more fashionable and more interesting theories of climatic change should be mentioned. The first involves a change in celestial mechanical variables: the shape of the Earth's orbit, the tilt of its axis of rotation, and the precession of that axis all vary over long periods of time because of the interaction of the Earth with other nearby celestial objects. Detailed calculations of the extent of such variations show that they can be responsible for at least a few degrees of temperature variation, and with the possibility of positive feedbacks this might, by itself, be adequate to explain major climatic variations.

A second class of theories involves albedo variations. One of the more striking causes for such variations is the injection into the Earth's atmosphere of massive amounts of dust—for example, from a volcanic explosion such as Krakatoa's in 1883. While there has been some debate on whether such dust heats or cools the Earth, the bulk of present calculations shows that the fine particulates, very slowly falling out of Earth's stratosphere, increase the Earth's albedo and therefore cool it. There is recent sedimentological evidence that past epochs of extensive production of volcanic particulates correspond in time to past epochs of glaciation and low temperatures. In addition, episodes of mountain building and the creation of land surface on the Earth increase the global albedo because the land is brighter than the water.

Finally, there is the possibility of variations in the brightness of the Sun. We know—from theories of solar evolution—that over many billions of years the Sun has been getting steadily brighter. This immediately poses a problem for the most ancient climatology of the Earth, because the Sun should have been 30 or 40 percent dimmer some 3 or 4 billion years ago; and this is enough, even with the greenhouse effect, to have resulted in global temperatures well below the freezing point of seawater. Yet there is extensive geological evidence—for example, underwater ripple marks, pillow lavas produced by the quenching of magma in the ocean, and fossil stromatolites produced by oceanic algae—that there was ample water then available. One proposed way out of this quandary is the possibility that there were additional greenhouse gases in the early atmosphere of the Earth—especially ammonia—which produced the required temperature increment. But apart from this very slow evolution of the brightness of the Sun, is it possible that shorter-term fluctuations occur? This is an important and unsolved problem, but recent difficulties in finding neutrinos—which should, according to current theories, be emitted from the interior of the Sun—have led to the suggestion that the Sun is today in an anomalously dim period.

The inability to distinguish between the various alternative models of climatic change might appear to be nothing more than an unusually annoying intellectual problem—except for the fact that there appear to be certain practical and immediate consequences of climatic change. Some evidence on the trend of global temperature seems to show a very slow increase from the beginning of the industrial revolution to about 1940, and an alarmingly steep decline in global temperature thereafter. This pattern has been attributed to the burning of fossil fuels, which has two consequences—the liberation of carbon dioxide, a greenhouse gas, into the atmosphere, and the simultaneous injection into the atmosphere of fine particles, from the incomplete burning of the fuel. The carbon dioxide heats the Earth; the fine particles, through their higher albedo, cool it. It may be that until 1940 the greenhouse effect was winning, and since then the increased albedo is winning.

The ominous possibility that human activities may cause inadvertent climate modification makes the interest in planetary climatology rather important. There are worrisome positive feedback possibilities on a planet with declining temperatures. For example, an increased burning of fossil fuels in a short-term attempt to stay warm can result in more rapid long-term cooling. We live on a planet in which agricultural technology is responsible for the food of more than a billion people. The crops have not been bred for hardiness against climatic variations. Human beings can no longer undertake great migrations in response to climatic change, or at least it is more difficult on a planet controlled by nation-states. It is becoming imperative to understand the causes of climatic variations and to develop the possibility of performing climatic re-engineering of the Earth.

Oddly enough, some of the most interesting hints on the nature of such climatic changes appear to be coming from studies not of the Earth at all, but of Mars. Mariner 9 was injected into Martian orbit on November

14, 1971. It had a useful scientific lifetime of a full terrestrial year and procured 7,200 photographs, covering the planet from pole to pole, as well as tens of thousands of spectra and other scientific information. As we saw earlier, when Mariner 9 arrived at Mars there was virtually no detail whatever to be seen on the surface because the planet was in the throes of a great global dust storm. It was readily observed that the atmospheric temperatures increased, but the surface temperatures decreased during the dust storm, and this simple observation immediately provides at least one clear case of the cooling of a planet by the massive injection of dust into its atmosphere. Calculations have been performed that use precisely the same physics for both the Earth and Mars and treat them as two different examples of the general problem of the climatic effects of massive dust injection into a planetary atmosphere.

There was another and entirely unexpected climatological finding by Mariner 9—the discovery of numerous sinuous channels, replete with tributaries, covering the equatorial and mid-latitudes of Mars. In all cases where relevant data exist, the channels are going in the proper direction—downhill. Some of them show braided patterns, sand bars, slumping of the banks, streamlined teardrop-shaped interior "islands" and other characteristic morphological signs of terrestrial river valleys.

But there is a great problem with the interpretation of the Martian channels as dry riverbeds, or arroyos: liquid water apparently cannot exist on Mars today. The pressures are simply too low. Carbon dioxide on Earth is known as both a solid and a gas, but never as a liquid (except in high-pressure storage tanks). In the same way, water on Mars can exist as a solid (ice or snow) or as vapor, but not as a liquid. For this reason some geologists are reluctant to accept the theory that at one time the channels contained liquid water. Yet they are dead ringers for terrestrial rivers, and at least many of them have forms inconsistent with other possible structures such as collapsed lava tubes, which may be responsible for sinuous valleys on the Moon.

Furthermore, there is an apparent concentration of such channels toward the Martian equator. The one striking fact about the equatorial regions of Mars is that they are the only places on the planet where the average daytime temperature is above the freezing point of water. And no other liquid is simultaneously cosmically abundant, of low viscosity, and with a freezing point below Martian equatorial temperatures.

If, then, the channels were made by running water on Mars, that water apparently must have run at a time when the Martian environment was significantly different from what it is today. Today Mars has a thin atmosphere, low temperatures and no liquid water. At some time in the past, it may have had higher pressures, perhaps somewhat higher temperatures and extensive running water. Such an environment appears to be more hospitable to forms of life based on familiar terrestrial biochemical principles than the present Martian environment.

A detailed study of the possible causes of such major climatic changes on Mars has laid stress on a feedback mechanism known as advective instability. The Martian atmosphere is composed primarily of carbon dioxide. There seem to be large repositories of frozen CO_2 in at least one of the two polar caps. The pressure of CO_2 in the Martian atmosphere is quite close to the pressure of CO_2 expected in equilibrium with frozen carbon dioxide at the temperature of the cold Martian pole. This is a situation quite similar to the pressure in a laboratory vacuum system determined by the temperature of a "cold finger" in the system. At the present time the Martian atmosphere is so thin that hot air, rising from the equator and settling at the poles, plays a very small role in heating the high latitudes. But let us imagine that the temperature in the polar regions is somehow slightly increased. The total atmospheric pressure increases, the efficiency of heat transport by advection from equator to pole also increases, polar temperatures increase still further, and we see the possibility of a runaway to high temperatures. Likewise a decrease in temperature, from whatever cause, could bring about a runaway toward a lower temperature. The physics of this Martian situation is easier to work out than the comparable case on Earth, because on Earth the major atmospheric constituents, oxygen and nitrogen, are not condensable at the poles.

For a major increase in pressure to occur on Mars, the amount of heat absorbed in the polar regions of the planet must be increased by some 15 or 20 percent for a period of at least a century. Three possible sources of variation in the heating of the cap have been identified, and they are, interestingly enough, very similar to the three fashionable models of terrestrial climatic change discussed above. In the first, variations of the tilt of the Martian rotational axis toward the Sun are invoked. Such variations are much more striking than for the Earth, because Mars is close to Jupiter, the most massive planet in the solar system, and the gravitational perturbations by Jupiter are pronounced. Here variations in global pressure and temperature will occur on hundred thousand to million year time scales.

Secondly, a variation in the albedo of the polar regions can cause major climatic variations. We can already see substantial sand and dust storms on Mars, because of which the polar caps seasonally darken and brighten. There has been one suggestion that the climate of Mars may be made more hospitable if a hardy species of polar plant can be developed that will lower the albedo of the Martian polar regions.

Finally, there is the possibility of variations in the luminosity of the Sun. Some of the channels on Mars have an occasional impact crater in them, and crude dating of the channels from the frequency of impacts from interplanetary space shows that some of them must be about a billion years old. This is reminiscent of the last epoch of high global temperatures on the planet Earth and raises the captivating possibility of synchronous major variations in climate between the Earth and Mars.

The subsequent Viking missions to Mars have increased our knowledge about the channels in a major way, have provided quite independent evidence for a dense earlier atmosphere and have demonstrated a great repository of frozen carbon dioxide in the polar ice. When the Viking results are fully assimilated, they promise to add greatly to our knowledge of the present environment as well as the past history of the planet, and of the comparison between the climates of the Earth and Mars.

When scientists are faced with extremely difficult theoretical problems, there is always the possibility of performing experiments. In studies of the climate of an entire planet, however, experiments are expensive and difficult to perform, and have potentially awkward social consequences. By the greatest good fortune, nature has come to our aid by providing us with nearby planets with significantly different climates and significantly different physical variables. Perhaps the sharpest test of theories of climatology is that they be able to explain the climates of all the nearby planets, Earth, Mars and Venus. Insights gained from the study of one planet will inevitably aid the study of the others. Comparative planetary climatology appears to be a discipline, just in the process of birth, with major intellectual interest and practical applications.

CHAPTER 15

KALLIOPE AND THE KAABA

◆◆◆

We imagine them
flitting
cheek to jowl,
these driftrocks
of cosmic ash
thousandfold afloat
between Jupiter and Mars.
Frigga,
Fanny,
Adelheid
Lacrimosa.
Names to conjure with,
Dakotan black hills,
a light-opera
staged on a barrier reef.
And swarm they may have,
crumbly as blue-cheese,
that ur-moment
when the solar system
broke wind.
But now
they lumber
so wide apart

from each
to its neighbor's
pinprick-glow
slant millions
and millions
of watertight miles.
Only in the longest view
do they graze
like one herd
on a breathless tundra.

DIANE ACKERMAN,
The Planets (New York, Morrow, 1976)

ONE OF THE seven wonders of the ancient world was the Temple of Diana at Ephesus, in Asia Minor, an exquisite example of Greek monumental architecture. The Holy of Holies in this temple was a great black rock, probably metallic, that had fallen from the skies, a sign from the gods, perhaps an arrowhead shot from the crescent moon, the symbol of Diana the Huntress.

Not many centuries later—perhaps even at the same time—another great black rock, according to the belief of many, fell out of the sky onto the Arabian Peninsula. There, in pre-Islamic times, it was emplaced in a Meccan temple, the Kaaba, and offered something akin to worship. Then, in the seventh and eight centuries A.D., came the stunning success of Islam, founded by Muhammed, who lived out most of his days not far from this large dark stone, the presence of which might conceivably have influenced his choice of career. The earlier worship of the stone was incorporated into Islam, and today a principal focus of every pilgrimage to Mecca is that same stone—often called the Kaaba after the temple that enshrines it. (All religions have shamelessly coopted their predecessors—e.g., consider the Christian festival of Easter, where the ancient fertility rites of the spring equinox are today cunningly disguised as eggs and baby animals. Indeed the very name Easter is, according to some etymologies, a corruption of the name of the great Near Eastern Earth mother goddess, Astarte. The Diana of Ephesus is a later and Hellenized version of Astarte and Cybelle.)

In primitive times, a great boulder falling out of a clear blue sky must have provided onlookers with a memorable experience. But it had a greater importance: at the dawn of metallurgy, iron from the skies was, in many parts of the world, the purest available form of this metal. The military significance of iron swords and the agricultural significance of iron plowshares made metal from the sky a concern of practical men.

Rocks still fall from the skies; farmers still occasionally break their plows on them; museums still pay a bounty for them; and, very rarely, one falls through the eaves of a house, narrowly missing a family in its evening hypnogogic ritual before the television set. We call these objects meteorites. But naming them is not the same as understanding them. Where, in fact, do meteorites come from?

Between the orbits of Mars and Jupiter are thousands of irregularly shaped, tumbling little worlds called asteroids or planetoids. "Asteroid" is not a good term for them because they are not like stars. "Planetoid" is much better because they *are* like planets, only smaller, but "asteroid" is the more widely used term by far. Ceres, the first asteroid to be found, was discovered* telescopically on January 1, 1801—an auspicious finding on the first day of the nineteenth century—by G. Piazzi, an Italian monk. Ceres is about 1,000 kilometers in diameter and is by far the largest asteroid. (By comparison, the diameter of the Moon is 3,464 kilometers.) Since then, more than two thousand asteroids have been discovered. Asteroids are given a number indicating

* Unexpected discoveries are useful for calibrating pre-existing ideas. G. W. F. Hegel has had a very powerful imprint on professional philosophy of the nineteenth and early twentieth centuries and a profound influence on the future of the world because Karl Marx took him very seriously (although sympathetic critics have argued that Marx's arguments would have been more compelling had he never heard of Hegel). In 1799 or 1800 Hegel confidently stated, using presumably the full armamentarium of philosophy available to him, that no new celestial objects could exist within the solar system. One year later, the asteroid Ceres was discovered. Hegel then seems to have returned to pursuits less amenable to disproof.

their order of discovery. But following Piazzi's lead, a great effort was also made to give them names—female names, preferably from Greek mythology. However, two thousand asteroids is a great many, and the nomenclature becomes a little ragged toward the end. We find 1 Ceres, 2 Pallas, 3 Juno, 4 Vesta, 16 Psyche, 22 Kalliope, 34 Circe, 55 Pandora, 80 Sappho, 232 Russia, 324 Bamberga, 433 Eros, 710 Gertrud, 739 Mandeville, 747 Winchester, 904 Rockefelleria, 916 America, 1121 Natasha, 1224 Fantasia, 1279 Uganda, 1556 Icarus, 1620 Geographos, 1685 Toro, and 694 Ekard (Drake [University] spelled backwards). 1984 Orwell is, unfortunately, a lost opportunity.

Many asteroids have orbits that are highly elliptical or stretched-out, not at all like the almost perfectly circular orbits of Earth or Venus. Some asteroids have their far points from the Sun beyond the orbit of Saturn; some have their near points to the Sun close to the orbit of Mercury; some, like 1685 Toro, live out their days between the orbits of Earth and Venus. Since there are so many asteroids on very elliptical orbits, collisions are inevitable over the lifetime of the solar system. Most collisions will be of the overtaking variety, one asteroid nudging up to another, making a soft splintering crash. Since the asteroids are so small, their gravity is low and the collision fragments will be splayed out into space into slightly different orbits from those of the parent asteroids. It can be calculated that such collisions will produce, on occasion, fragments that by accident intercept the Earth, fall through its atmosphere, survive the ablation of entry, and land at the feet of a quite properly astonished itinerant tribesman.

The few meteorites that have been tracked as they enter the Earth's atmosphere originated back in the main asteroid belt, between Mars and Jupiter. Laboratory studies of the physical properties of some meteorites show them to have originated where the temperatures are those of the main asteroid belt. The evidence is clear: the meteorites ensconced in our museums are fragments of asteroids. We have on our shelves pieces of cosmic objects!

But which meteorites come from which asteroids? Until the last few years, answering this question was beyond the powers of planetary scientists. Recently, however, it has become possible to perform spectrophotometry of asteroids in visible and near-infrared radiation; to examine the polarization of sunlight reflected off asteroids as the geometry of the asteroid, the Sun and Earth changes; and to examine the middle-infrared emission of the asteroids. These asteroid observations, and comparable studies of meteorites and other minerals in the laboratory, have provided the first fascinating hints on the correlation between specific asteroids and specific meteorites. More than 90 percent of the asteroids studied fall into one of two composition groups: stony-iron or carbonaceous. Only a few percent of the meteorites on Earth are carbonaceous, but carbonaceous meteorites are very friable and rapidly weather to powder under typical terrestrial conditions. They probably also fragment more readily upon entry into the Earth's atmosphere. Since stony-iron meteorites are much hardier, they are disproportionately represented in our museum collections of meteorites. The carbonaceous meteorites are rich in organic compounds, including amino acids (the building blocks of proteins), and may be representative of the materials from which the solar system was formed some 4.6 billion years ago.

Among the asteroids which appear to be carbonaceous are 1 Ceres, 2 Pallas, 19 Fortuna, 324 Bamberga and 654 Zelinda. If asteroids that are carbonaceous on the outside are also carbonaceous on the inside, then most of the asteroidal material is carbonaceous. They are generally dark objects, reflecting only a small percent of the light shining on them. Recent evidence suggests that Phobos and Deimos, the two moons of Mars, may also be carbonaceous, and are perhaps carbonaceous asteroids that have been captured by Martian gravity.

Typical asteroids showing properties of stony-iron meteorites are 3 Juno, 8 Flora, 12 Victoria, 89 Julia and 433 Eros. Several asteroids fit into some other category: 4 Vesta resembles a kind of meteorite called a basaltic achondrite, while 16 Psyche and 22 Kalliope appear to be largely iron.

The iron asteroids are interesting because geophysicists believe that the parent body of an object greatly enriched in iron must have been molten so as to differentiate, to separate out the iron from the silicates in the initial chaotic jumble of the elements in primordial times. On the other hand, for the organic molecules in carbonaceous meteorites to have survived at all they must never have been raised to temperatures hot enough to melt rock or iron. Thus, different histories are implied for different asteroids.

From the comparison of asteroidal and meteoritic properties, from laboratory studies of meteorites and computer projections back in time of asteroidal motions, it may one day be possible to reconstruct asteroid histories. Today we do not even know whether they represent a planet that was prevented from forming because of the powerful gravitational perturbations of nearby Jupiter, or whether they are the remnants of a fully formed planet that somehow exploded. Most students of the subject incline to the former hypothesis because no one can figure out how to blow up a planet—which is just as well. Eventually we may be able to piece together the whole story.

There may also be in hand meteorites which do not come from asteroids. Perhaps there are fragments of young comets, or of the moons of Mars, or of the surface of Mercury, or of the satellites of Jupiter, sitting dusty and ignored in some obscure museum. But it is clear that the true picture of the origin of the meteorites is beginning to emerge.

The Holy of Holies in the Temple of Diana at Ephesus has been destroyed. But the Kaaba has been carefully preserved, although there seems never to have been a true scientific examination of it. There are some who believe it to be a dark, stony rather than metallic meteorite. Recently two geologists have suggested, on admittedly quite fragmentary evidence, that it is instead an agate. Some Muslim writers believe that the color of the Kaaba was originally white, not black, and that the present color is due to its repeated handling. The official view of the Keeper of the Black Stone is that it was placed in its present position by the patriarch Abraham and fell from a religious rather than an astronomical heaven—so that no conceivable physical test of the object could be a test of Islamic doctrine. It would nevertheless be of great interest to examine, with the full armory of modern laboratory techniques, a small fragment of the Kaaba. Its composition could be determined with precision. If it is a meteorite, its cosmic-ray-exposure age—the time spent from fragmentation to arrival on Earth—could be established. And it would be possible to test hypotheses of origin: such as, for example, the idea that some 5 million years ago, about the time of the origin of the hominids, the Kaaba was chipped off an asteroid named 22 Kalliope, orbited the Sun for ages of geological time, and then accidentally encountered the Arabian Peninsula 2,500 years ago.

CHAPTER 16

THE GOLDEN AGE OF PLANETARY EXPLORATION

◆◆◆

The unquiet republic of the maze
Of Planets, struggling fierce towards heaven's free
wilderness.

PERCY BYSSHE SHELLEY,
Prometheus Unbound **(1820)**

MUCH OF HUMAN HISTORY can, I think, be described as a gradual and sometimes painful liberation from provincialism, the emerging awareness that there is more to the world than was generally believed by our ancestors. With awesome ethnocentrism, tribes all over the Earth called themselves "the people" or "all men," relegating other groups of humans with comparable accomplishments to subhuman status. The high civilization of ancient Greece divided the human community into Hellenes and barbarians, the latter named after an uncharitable imitation of the languages of non-Greeks ("Bar Bar . . ."). That same classical civilization, which in so many respects is the antecedent of our own, called its small inland sea the Mediterranean—which means the middle of the Earth. For thousands of years China called itself the Middle Kingdom, and the meaning was the same: China was at the center of the universe and the barbarians lived in outer darkness.

Such views or their equivalent are only slowly changing, and it is possible to see some of the roots of racism and nationalism in their pervasive early acceptance by virtually all human communities. But we live in an extraordinary time, when technological advances and cultural relativism have made such ethnocentrism much more difficult to sustain. The view is emerging that we all share a common life raft in a cosmic ocean, that the Earth is, after all, a small place with limited resources, that our technology has now attained such powers that we are able to affect profoundly the environment of our tiny planet. This deprovincialization of mankind has been aided powerfully, I believe, by space exploration—by exquisite photographs of the Earth taken from a great distance, showing a cloudy, blue, spinning ball set like a sapphire in the endless velvet of space; but also by the exploration of other worlds, which have revealed both their similarities and their differences to this home of mankind.

We still talk of "the" world, as if there were no others, just as we talk about "the" Sun and "the" Moon. But there are many others. Every star in the sky is a sun. The rings of Uranus represent millions of previously unsuspected satellites orbiting Uranus, the seventh planet. And, as space vehicles have demonstrated so dramatically in the last decade and a half, there are other worlds—nearby, relatively accessible, profoundly interesting, and not a one closely similar to ours. As these planetary differences, and the Darwinian insight that life elsewhere is likely to be fundamentally different from life here, become more generally perceived, I believe they will provide a cohesive and unifying influence on the human family, which inhabits, for a time, this unprepossessing world among an immensity of others.

Planetary exploration has many virtues. It permits us to refine insights derived from such Earth-bound sciences as meteorology, climatology, geology and biology, to broaden their powers and improve their practical applications here on Earth. It provides cautionary tales on the alternative fates of worlds. It is an aperture to future high technologies important for life here on Earth. It provides an outlet for the traditional human zest for exploration and discovery, our passion to find out, which has been to a very large degree responsible for our success as a species. And it permits us, for the first time in history, to approach with rigor, with a significant chance of finding out the true answers, questions on the origins and destinies of worlds, the beginnings and ends of life, and the possibility of other beings who live in the skies—questions as basic to the human enterprise as thinking is, as natural as breathing.

Interplanetary unmanned spacecraft of the modern generation extend the human presence to bizarre and exotic landscapes far stranger than any in myth or legend. Propelled to escape velocity near the Earth, they adjust their trajectories with small rocket motors and tiny puffs of gas. They power themselves with sunlight and with nuclear energy. Some take only a few days to traverse the lake of space between Earth and Moon; others may take a year to Mars, four years to Saturn, or a decade to traverse the inland sea between us and distant Uranus. They float serenely on pathways predetermined by Newtonian gravitation and rocket technology, their bright metal gleaming, awash in the sunlight which fills the spaces between the worlds. When they arrive at their destinations, some will fly by, garnering a brief glimpse of an alien planet, perhaps with a retinue of moons, before continuing on farther into the depths of space. Others insert themselves into orbit about another world to examine it at close range, perhaps for years, before some essential component runs down or wears out. Some spacecraft will make landfall on another world, decelerating by atmospheric friction or parachute drag or the precision firing of retrorockets before gently setting down somewhere else. Some landers are stationary, condemned to examine a single spot on a world awaiting exploration. Others are self-propelled, slowly wandering to a distant horizon which holds no man knows what. And still others are capable of remotely acquiring rock and soil—a sample of another world—and returning it to the Earth.

All these spacecraft have sensors that extend astonishingly the range of human perception. There are devices that can determine the distribution of radioactivity over another planet from orbit; that can feel from the surface the faint rumble of a distant planetquake deep below; that can obtain three-dimensional color or infrared images of a landscape like none ever seen on Earth. These machines are, at least to a limited degree, intelligent. They can make choices on the basis of information they themselves receive. They can remember with great accuracy a detailed set of instructions which, if written out in English, would fill a good-sized book. They are obedient and can be reinstructed by radio messages sent to them from human controllers on Earth. And they have returned, mostly by radio, a rich and varied harvest of information on the nature of the solar system we inhabit. There have been fly-bys, crash-landers, soft-landers, orbiters, automated roving vehicles, and unmanned returned sample missions from our nearest celestial neighbor, the Moon—as well as, of course, six successful and heroic manned expeditions in the Apollo series. There has been a fly-by of Mercury; orbiters, entry probes and landers on Venus; fly-bys, orbiters and landers to Mars; and fly-bys of Jupiter and Saturn. Phobos and Deimos, the two small moons of Mars, have been examined close up, and tantalizing images have been obtained of a few of the moons of Jupiter.

We have caught our first glimpses of the ammonia clouds and great storm systems of Jupiter; the cold, salt-covered surface of its moon, Io; the desolate, crater-pocked, ancient and broiling Mercurian wasteland; and the wild and eerie landscape of our nearest planetary neighbor, Venus, where the clouds are composed of an acid rain that falls continuously but never patters the surface because that hilly landscape, illuminated by sunlight diffusing through the perpetual cloud layer, is everywhere at 900°F. And Mars: What a puzzle, what a joy, enigma and delight is Mars, with ancient river bottoms; immense, sculpted polar terraces; a volcano almost 80,000 feet high; raging windstorms; balmy afternoons; and an apparent initial defeat of our first pioneering effort to answer the question of questions—whether the planet harbors, now or ever, a home-grown form of life.

There are on Earth only two spacefaring nations, only two powers so far able to send machines much beyond the Earth's atmosphere—the United States and the Soviet Union. The United States has accomplished the only manned missions to another body, the only successful Mars landers and the only expeditions to Mercury, Jupiter and Saturn. The Soviet Union has pioneered the automated exploration of the Moon, including the only unmanned rovers and return sample missions on any celestial objects, and the first entry probes and landers on Venus. Since the end of the Apollo program, Venus and the Moon have become, to a certain degree, Russian turf, and the rest of the solar system visited only by American space vehicles. While there is a certain degree of scientific cooperation between the two spacefaring nations, this planetary territoriality has come about by default rather than by agreement. There have in recent years been a set of very ambitious but unsuccessful Soviet missions to Mars, and the United States launched a modest but successful set of Venus orbiters and entry probes in 1978. The solar system is very large and there is much to explore. Even tiny Mars has a surface area comparable to the land area of the Earth. For practical reasons it is much easier to organize separate but coordinated missions launched by two or more nations than cooperative multinational ventures. In the sixteenth and seventeenth centuries, England, France, Spain, Portugal and Holland each organized on a grand scale missions of global exploration and discovery in vigorous competition. But the economic and religious motives of exploratory competition then do not seem to have their counterparts today. And there is every reason to think that national competition in the exploration of the planets will, at least for the foreseeable future, be peaceful.

THE LEAD TIMES for planetary missions are very long. The design, fabrication, testing, integration and launch of a typical planetary mission takes many years. A systematic program of planetary exploration requires a continuing commitment. The most celebrated American achievements on the Moon and planets—Apollo, Pioneer, Mariner and Viking—were initiated in the 1960s. At least until recently, the United States has made only one major commitment to planetary exploration in the whole of the decade of the 1970s—the Voyager missions, launched in the summer of 1977, to make the first systematic fly-by examination of Jupiter, Saturn, their twenty-five or so moons and the spectacular rings of the latter.

This absence of new starts has produced a real crisis in the community of American scientists and engineers responsible for the succession of engineering successes and high scientific discovery that began in 1962 with the Mariner 2 fly-by of Venus. There has been an interruption in the pace of exploration. Workers have been laid off and drifted to quite different jobs, and there is a real problem in providing continuity to the next generation of planetary exploration. For example, the earliest likely response to the spectacularly successful and historic Viking exploration of Mars will be a mission that does not even arrive at the Red Planet before 1985—a gap in Martian exploration of almost a decade. And there is not the slightest guarantee that there will be a mission even then. This trend—a little like dismissing most of the shipwrights, sail weavers and navigators of Spain in the early sixteenth century—shows some slight signs of reversal. Recently approved was Project Galileo, a middle-1980s mission to perform the first orbital reconnaissance of Jupiter and to drop the first probe into its atmosphere—which may contain organic molecules synthesized in a manner analogous to the chemical events which on Earth led to the origin of life. But the following year Congress so reduced the funds available for Galileo that it is, at the present writing, teetering on the brink of disaster.

In recent years the entire NASA budget has been well below one percent of the federal budget. The funds spent on planetary exploration have been less than 15 percent of that. Requests by the planetary science community for new missions have been repeatedly rejected —as one senator explained to me, the public has not, despite *Star Wars* and *Star Trek,* written to Congress in support of planetary missions, and scientists do not constitute a powerful lobby. And yet, there are a set of missions on the horizon that combine extraordinary scientific opportunity with remarkable popular appeal:

Solar Sailing and Comet Rendezvous. In ordinary interplanetary missions, spacecraft are obliged to follow trajectories that require a minimum expenditure of energy. The rockets burn for short periods of time in the vicinity of Earth, and the spacecraft mainly coast for the rest of the journey. We have done as well as we have not because of enormous booster capability, but because of great skill with severely constrained systems. As a result, we must accept small payloads, long mission times and little choice of departure or arrival dates. But just as on Earth we are considering moving from fossil fuels to solar power, so it is in space. Sunlight exerts a small but palpable force called radiation pressure. A sail-like structure with a very large area for its mass can use radiation pressure for propulsion. By positioning the sail properly, we can be carried by sunlight both inwards toward and outwards away from the Sun. With a square sail about half a mile on each side, but thinner than the thinnest Mylar, interplanetary missions can be accomplished more efficiently than with conventional rocket propulsion. The sail would be launched into Earth orbit by the manned Shuttle craft, unfurled and strutted. It would be an extraordinary sight, easily visible to the naked eye as a bright point of light. With a pair of binoculars, detail on such a sail could be made out—perhaps even what on seventeenth-century sailing ships was called the "device," some appropriate graphic symbol, perhaps a representation of the planet Earth. Attached to the sail would be a scientific spacecraft designed for a particular application.

One of the first and most exciting applications being discussed is a comet-rendezvous mission, perhaps a rendezvous with Halley's comet in 1986. Comets spend most of their time in interstellar space and should provide major clues on the early history of the solar system and the nature of the matter between the stars. Solar sailing to Halley's comet might not only provide close-up pictures of the interior of a comet—about which we now know close to nothing—but also, astonishingly, return a piece of a comet to the planet Earth. The practical advantages and the romance of solar sailing are both evident in this example, and it is clear that it represents not just a new mission but a new interplanetary technology. Because the development of solar-sailing technology is behind that of ion propulsion, it is the latter that may propel us on our first missions to the comets. Both propulsion mechanisms have their place in future interplanetary travel. But in the long term I believe solar sailing will make the greater impact. Perhaps by the early twenty-first century there will be interplanetary regattas competing for the fastest time from Earth to Mars.

Mars Rovers. Before the Viking mission, no terrestrial spacecraft had successfully landed on Mars. There had been several Soviet failures, including at least one which was quite mysterious and possibly attributable to the hazardous nature of the Martian landscape. Thus, both Viking 1 and Viking 2 were, after painstaking efforts, successfully landed in two of the dullest places we could find on the Martian surface. The lander stereo cameras showed distant valleys and other inaccessible vistas. The orbital cameras showed an extraordinarily varied and geologically exuberant landscape which we could not examine close up with the stationary Viking lander. Further Martian exploration, both geological and biological, cries out for roving vehicles capable of landing in the safe but dull places and wandering hundreds or thousands of kilometers to the exciting places. Such a rover would be able to wander to its own horizon every day and produce a continuous stream of photographs of new landscapes, new phenomena and very likely major surprises on Mars. Its importance would be improved still further if it operated in tandem with a Mars polar orbiter which would geochemically map the planet, or with an unmanned Martian aircraft which would photograph the surface from very low altitudes.

Titan Lander. Titan is the largest moon of Saturn and the largest satellite in the solar system (see Chapter 13). It is remarkable for having an atmosphere denser than that of Mars and is probably covered with a layer of brownish clouds composed of organic molecules. Unlike Jupiter and Saturn, it has a surface on which we can land, and its deep atmosphere is not so hot as to destroy the organic molecules. A Titan entry-probe and lander mission would probably be part of a Saturn orbital mission, which might also include a Saturn entry probe.

Venus Orbital Imaging Radar. The Soviet Venera 9 and 10 missions have returned the first close-up photographs of the surface of Venus. Because of the permanent cloud pall, the surface features of Venus are not visible through Earth-bound optical telescopes. However, Earth-based radar and the radar system aboard the small Pioneer Venus orbiter have now begun to map Venus surface features, and have revealed mountains and craters and volcanoes as well as stranger morphology. A proposed Venus orbital imaging radar would provide pole-to-pole radar pictures of Venus with much higher detail than can be achieved from the surface of the Earth, and would permit a preliminary reconnaissance of the Venus surface comparable to that achieved for Mars in 1971–72 by Mariner 9.

Solar Probe. The Sun is the nearest star, the only one we are likely to be able to examine close up, at least for many decades. A near approach to the Sun would be of great interest, would help in understanding its influence on Earth, and would also provide vital additional tests of such theories of gravitation as Einstein's General Theory of Relativity. A solar probe mission is difficult for two reasons: the energy required to undo the Earth's (and the probe's) motion around the Sun so it can fall into the Sun, and the intolerable heating as the probe approaches the Sun. The first problem can be solved by launching the spacecraft out to Jupiter and then using Jupiter's gravitation to fling it into the Sun. Since there are many asteroids interior to Jupiter's orbit, this might possibly be a useful mission for studying asteroids as well. An approach to the second problem, at first sight remarkable for its naïveté, is to fly into the Sun *at night*. On Earth, nighttime is of course merely the interposition of the solid body of the Earth between us and the Sun. Likewise for a solar probe. There are some asteroids that come rather close to the Sun. A solar probe would approach the Sun in the shadow of a Sun-grazing asteroid (meanwhile making observations of the asteroid as well). Near the point of closest approach of the asteroid to the Sun, the probe would emerge from the asteroidal shadow and plunge, filled with a fluid that resists heating, as deeply into the atmosphere of the Sun as it could until it melted and vaporized—atoms from the Earth added to the nearest star.

Manned Missions. As a rule of thumb, a manned mission costs from fifty to a hundred times more than a comparable unmanned mission. Thus, for scientific exploration alone, unmanned missions, employing machine intelligence, are preferred. However, there may well be reasons other than scientific for exploring space —social, economic, political, cultural or historical. The manned missions most frequently talked about are space stations orbiting the Earth (and perhaps devoted to harvesting sunlight and transmitting it in microwave beams down to an energy-starved Earth), and a permanent lunar base. Also being discussed are rather grand schemes for the construction of permanent space cities in Earth orbit, constructed from lunar or asteroidal materials. The cost of transporting materials from such low-gravity worlds as the Moon or an asteroid to Earth orbit is much less than transporting the same materials from our high-gravity planet. Such space cities might ultimately be self-propagating—new ones constructed by older ones. The costs of these large manned stations have not yet been estimated reliably, but it seems likely that all of them—as well as a manned mission to Mars—would cost in the $100 billion to $200 billion range. Perhaps such schemes will one day be implemented; there is much that is far-reaching and historically significant in them. But those of us who have fought for years to organize space ventures costing less than one percent as much may be forgiven for wondering whether the required funds will be allocated, and whether such expenditures are socially responsible.

However, for substantially less money, an important expedition that is preparatory for each of these manned ventures could be mustered—an expedition to an Earth-crossing carbonaceous asteroid. The asteroids occur mostly between the orbits of Mars and Jupiter. A small fraction of them have trajectories that carry them across Earth's orbit and occasionally within a few million miles of the Earth. Many asteroids are mainly carbonaceous—with large quantities of organic materials and chemically bound water. The organic matter is thought to have condensed in the very earliest stages of the formation of the solar system from interstellar gas and dust, some 4.6 billion years ago, and their study and comparison with cometary samples would be of extraordinary scientific interest. I do not think that materials from a carbonaceous asteroid are likely to be criticized in the same way that the Apollo returned lunar samples were—as being "only" rocks. Moreover, a manned landing on such an object would be an excellent preparation for the eventual exploitation of resources in space. And finally, landing on such an object would be fun: because the gravity field is so low, it would be possible for an astronaut to do a standing high jump of about ten kilometers. These Earth-crossing objects, which are being discovered at a rapidly increasing pace, are called—by a name selected long before manned spaceflight—the Apollo objects. They may or may not be the dead husks of comets. But whatever their origin, they are of great interest. Some of them are the easiest objects in space for humans to get to, using only the

Shuttle technology, which will be available in another few years.

THE SORTS of missions I have outlined are well within our technological capability and require a NASA budget not much larger than the present one. They combine scientific and public interest, which very often share coincident objectives. Were such a program carried out, we would have made a preliminary reconnaissance of all the planets and most of the moons from Mercury to Uranus, made a representative sampling of asteroids and comets, and discovered the boundaries and contents of our local swimming hole in space. As the finding of rings around Uranus reminds us, major and unexpected discoveries are waiting for us. Such a program would also have made the first halting steps in the utilization of the solar system by our species, tapping the resources on other worlds, arranging for human habitation in space, and ultimately reworking or terraforming the environments of other planets so that human beings can live there with minimal inconvenience. Human beings will have become a multi-planet species.

The transitional character of these few decades is evident. Unless we destroy ourselves, it is clear that humanity will never again be restricted to a single world. Indeed, the ultimate existence of cities in space and the presence of human colonies on other worlds will make it far more difficult for the human species to self-destruct. It is clear that we have entered, almost without noticing it, a golden age of planetary exploration. As in many comparable cases in human history, the opening of horizons through exploration is accompanied by an opening of artistic and cultural horizons. I do not imagine that many people in the fifteenth century ever wondered if they were living in the Italian Renaissance. But the hopefulness, the exhilaration, the opening of new ways of thought, the technological developments, the goods from abroad, and the deprovincialization of that age were then apparent to thoughtful men and women. We have the ability and the means and—I very much hope—the will for a comparable endeavor today.

For the first time in human history, it is within the power of this generation to extend the human presence to the other worlds of the solar system—with awe for their wonders, and a thirst for what they have to teach us.

PART IV
THE FUTURE

CHAPTER 17

"WILL YOU WALK A LITTLE FASTER?"

♦♦♦

"Will you walk a little faster?" said a whiting
to a snail,
"There's a porpoise close behind us,
and he's treading on my tail."

LEWIS CARROLL,
Alice in Wonderland

FOR MUCH OF human history we could travel only as fast as our legs would take us—for any sustained journey, only a few miles an hour. Great journeys were undertaken, but very slowly. For example, 20,000 or 30,000 years ago, human beings crossed the Bering Strait and for the first time entered the Americas, gradually working their way down to the southernmost tip of South America, in Tierra del Fuego, where Charles Darwin encountered them on the memorable voyage of H.M.S. *Beagle.* A concerted and single-minded effort of a dedicated band to walk from the straits between Asia and Alaska to Tierra del Fuego might have succeeded in a matter of years; in fact, it probably took thousands of years for diffusion of the human population to carry it so far south.

The original motivation for traveling fast must have

been, as the whiting's plaint reminds us, to escape from enemies and predators, or else to seek enemies and prey. A few thousand years ago a remarkable discovery was made: the horse can be domesticated and ridden. The idea is a very peculiar one, the horse not having been evolved for humans to ride. If looked at objectively, it is only a little less silly than, say, an octopus riding a grouper. But it worked and—especially after the invention of the wheel and the chariot—horseback or horse-drawn vehicles represented for millennia the most advanced transportation technology available to the human species. One can travel as much as 10 or perhaps even 20 miles an hour with horse technology.

We have emerged from horse technology only very recently—as, for example, our use of the term "horsepower" to rate automobile engines clearly shows. An engine rated at 375 horsepower has very roughly the pulling capacity of 375 horses. A team of 375 horses would make a very interesting sight. Arrayed in ranks of five horses each, the team would extend for about two-tenths of a mile in length and would be astonishingly unwieldy. On many roads the front rank of horse would be out of sight of the driver. And, of course, 375 horses do not travel 375 times as fast as one horse. Even with enormous teams of horses the speed of transportation was only ten or so times faster than when we could depend upon only our legs.

Thus the changes of the last century in transportation technology are striking. We humans have relied on legs for millions of years; horses for thousands; the internal-combustion engine for less than a hundred; and rockets for transportation for a few decades. But these products of human inventive genius have enabled us to travel on the land and on the surface of the waters a hundred times faster than we can walk, in the air a thousand times faster, and in space more than ten thousand times faster.

It used to be that the speed of communication was the same as the speed of transportation. There were a few fast communication methods earlier in our history —for example, signal flags or smoke signals or even one or two attempts at arrays of signal towers with mirrors employed to reflect sunlight or moonlight from one to another. News of the recapture of the Fortress of Györ by Hungarian commandos from the Turks was apparently conveyed to the Hapsburg Emperor Rudolf II through such a device: the "moonbeam telegraph," invented by the English astrologer John Dee, which apparently consisted of ten relay stations placed at intervals of forty kilometers between Györ and Prague. But with only a few exceptions, these methods proved impractical, and communications proceeded no faster than a man or a horse. This is no longer true. Communication by telephone and radio is now at the velocity of light—186,000 miles per second, or about two-thirds of a billion miles per hour. This is not simply the latest advance: it is the last advance. So far as we know, from Einstein's Special Theory of Relativity, the universe is constructed in such a way (at least around here) that no material object and no information can be transmitted faster than the velocity of light. This is not an engineering barrier like the so-called sound barrier, but a fundamental cosmic speed limit built deeply into the fabric of nature. Still, two-thirds of a billion miles per hour is fast enough for most practical purposes.

What is remarkable is that in communications technology we have already reached this ultimate limit and have adapted to it so well. There are few people who emerge breathless and palpitating from a routine long-distance telephone call, astounded at the speed of transmission. We take this almost instantaneous means of communication for granted. Yet in transportation technology, while we have not achieved speeds at all approaching the velocity of light, we find ourselves colliding with other limits, physiological and technological:

Our planet turns. When it is midday at one spot on the Earth, it is the dead of night on the other side. The Earth has therefore been conveniently arranged into twenty-four time zones of more or less equal width, making strips of longitude around the planet. If we fly very fast, we create situations our minds can accommodate but our bodies can abide only with great difficulty. It is a commonplace today to fly in relatively short trips westward and arrive before we leave—for example, when we take less than an hour to fly between two points separated by one time zone. When I take a 9 P.M. flight to London, it is already tomorrow at my destination. When I arrive, after a five- or six-hour flight, it is late at night for me but the beginning of the business day at my destination. My body senses something wrong, my circadian rhythms go awry, and it takes a few days to get adjusted to English time. A flight from New York to New Delhi is, in this respect, even more vexing.

I find it very interesting that two of the most gifted and inventive science-fiction writers of the twentieth century—Isaac Asimov and Ray Bradbury—both refuse to fly. Their minds have come to grips with interplanetary and interstellar spaceflight, but their bodies rebel at a DC-3. The rate of change in transportation technology has simply been too great for many of us to accommodate conveniently.

Much stranger possibilities are now practical. The Earth turns on its axis once every twenty-four hours. The circumference of the Earth is 25,000 miles. Thus, if we were able to travel at 25,000/24 = 1,040 miles per hour, we could just compensate for the Earth's rotation, and traveling westward at sunset, could maintain ourselves at sunset for the entire journey even if we circumnavigated the planet. (In fact, such a journey would also maintain us at the same *local* time as we journey westward from time zone to time zone, until we cross the international dateline and plunge precipitously into tomorrow.) But 1,040 miles per hour is less than twice the speed of sound and there are, worldwide, dozens of kinds of aircraft, chiefly military, that are capable of such speeds.*

* In manned Earth orbital flights, still other problems arise. Consider a religious Muslim or Jew circling the Earth once every ninety minutes. Is he obligated to celebrate the Sabbath every seventh orbit? Spaceflight provides access to environments very different from those in which we and our customs have grown up.

Some commercial aircraft, such as the Anglo-French Concorde, have comparable capabilities. The question, I think, is not: Can we go faster? but Do we have to? There has been concern expressed, some of it in my view quite appropriately, about whether the conveniences supersonic transports provide can possibly compensate for their overall cost and their ecological impact.

Most of the demand for high-speed long-distance travel comes from businessmen and government officials who need to have conferences with their opposite numbers in other states or countries. But what is really involved here is not the transportation of material but the transportation of information. I think much of the necessity for high-speed transport could be avoided if the existing communications technology were better used. I have many times participated in government or private meetings in which there were, say, twenty participants, each of whom was paid $500 for transportation and living expenses merely to attend the meeting—the cost of which was therefore $10,000 just to get the participants together. But all the participants ever exchange is information. Video phones, leased telephone lines, and facsimile reproducers to transmit paper copies of notes and diagrams would, I believe, serve as well or even better. There is no significant function of such a meeting—including private discussions among the participants "in the corridor"—that cannot be performed less expensively and at least equally conveniently with communications rather than transportation technology.

There are certainly advances in transportation that seem to me promising and desirable: vertical takeoff and landing (VTOL) aircraft are a remarkable boon for isolated and remote communities in case of medical or other emergencies. But the recent advances in transportation technology that I find most appealing are rubber fins for snorkel and scuba diving and hang gliders. These are technological advances much in the spirit of those sought by Leonardo da Vinci in mankind's first serious technological pursuit of flight in the fifteenth century; they permit an individual human being with little more than his own resources to enter—at a speed that is adequately exhilarating—another medium entirely.

WITH THE DEPLETION of fossil fuels I think it very likely that automobiles powered by internal-combustion engines will be with us for at most a few decades longer. The transportation of the future will simply have to be different. We can imagine quite comfortable and adequately speedy steam, solar, fuel-cell or electric ground vehicles, generating very little pollution and employing a technology comfortably accessible to the user.

Many responsible medical experts are concerned that we in the West—and increasingly even in developing countries—are becoming too sedentary. Driving an automobile exercises very few muscles. The demise of the automobile surely has many positive aspects when viewed in the long run, one of which is a return to the oldest transportation mechanism, walking, and to bicycling, which is in many ways the most remarkable.

I can easily imagine a healthy and stable future society in which walking and bicycling are the primary means of transportation; with pollution-free low-speed ground cars and railed public transportation systems widely available, and the most sophisticated transportation devices used relatively rarely by the average person. The one application of transportation technology that requires the most sophisticated technology is spaceflight. The returns in immediate practical benefits, scientific knowledge and appealing exploration provided by unmanned spaceflight are very impressive, and I would expect an increasing rate of space-vehicle launches by many nations in the next few decades, using more subtle forms of transportation, as described in the previous chapter. Nuclear electric, solar sailing and ion propulsion schemes have been proposed and are to some degree under development. As nuclear-fusion power plants are developed for Earth-bound applications in a few decades, there should be a development of fusion space engines as well.

The gravitational forces of planets have already been used to give velocities otherwise unobtainable. Mariner

10 reached Mercury only because it flew so close to Venus that Venus' gravity provided a significant boost in speed. And Pioneer 10 was boosted into an orbit that will carry it out of the solar system entirely, only because of a close passage by the giant planet Jupiter. In a way Pioneer 10 and 11 and Voyager 1 and 2 are our most advanced transportation systems. They are leaving the solar system at a speed of roughly 43,000 miles per hour, carrying messages to anyone who may intercept them out there in the dark of the night sky from the people of the Earth—who, only a little while ago, could travel no faster than a few miles per hour.

CHAPTER 18

VIA CHERRY TREE, TO MARS

◆◆◆

O for a Muse of fire, that would ascend
The brightest heaven of invention . . .

WILLIAM SHAKESPEARE,
Henry V, Prologue

IT IS A LAZY afternoon in an exquisite New England autumn. In about ten weeks it will be January 1, 1900, and your diary, into which are committed the events and ideas of your young life, will never again bear an entry with a date in the 1800s. You have just turned seventeen. You are looking forward to being a sophomore in high school, but you are now at home, in part because your mother is seriously ill with tuberculosis and in part because of your own chronic stomach pains. You are bright, with a certain flair for the sciences, but no one has ever indicated that you might have an extraordinary talent. You are complacently viewing the New England countryside from the limb of a tall old cherry tree which you have climbed, when suddenly you are struck by an idea, an overpowering and compelling vision that it might be possible, in fact rather than in fancy, to voyage to the planet Mars.

When you descend from the cherry tree you know that you are a very different boy from the one who climbed it. Your life's work is clearly set out for you, and for the next forty-five years your dedication never wavers. You have been smitten by the vision of flight to the planets. You are deeply moved and quietly awed by the vision in the cherry tree. The next year, on the anniversary of that vision, you climb the tree again to savor the joy and meaning of the experience; and forever after you make a point in your diary of calling the anniversary of that experience "Anniversary Day"—every October 19 until your death in the middle 1940s, by which time your theoretical insights and practical innovations have solved essentially all technological impediments to interplanetary flight.

Four years after your death a WAC Corporal mounted on the nose of a V-2 is successfully fired to an altitude of 250 miles, for all practical purposes to the threshold of space. All essential design elements of the WAC Corporal and the V-2, and the very concept of the multiple staging of rockets, have been worked out by you. A quarter of a century later, unmanned space vehicles will have been launched to all the planets known to ancient man; a dozen men will have set foot on the Moon; and two exquisitely miniaturized spacecraft named Viking will be on their way to Mars to attempt the first search for life on that planet.

ROBERT H. GODDARD never questioned or equivocated on the resolve he made in the cherry tree on the farm of his great-aunt Czarina in Worcester, Massachusetts. While there were others who had comparable visions—notably Konstantin Eduardovich Tsiolkovsky in Russia—Goddard represented a unique combination of visionary dedication and technological brilliance. He studied physics because he needed physics to get to Mars. He was for many years professor of physics and chairman of the physics department at Clark University in his hometown of Worcester.

In reading the notebooks of Robert Goddard, I am struck by how powerful his exploratory and scientific motivations were, and how influential speculative ideas —even erroneous ones—can be on the shaping of the future. In the few years surrounding the turn of the century, Goddard's interests were profoundly influenced by the idea of life on other worlds. He was intrigued by the claims of W. H. Pickering, of the Harvard College Observatory, that the Moon has a perceptible atmosphere, active volcanism, variable frost patches, and even changing dark markings, which Pickering interpreted variously as the growth of vegetation or even as the migration of enormous insects across the floor of the crater Eratosthenes. Goddard was captivated by the science fiction of H. G. Wells and Garrett P. Serviss, particularly the latter's *Edison's Conquest of Mars,* which, Goddard reported, "gripped my imagination tremendously." He attended and enjoyed lectures by Percival Lowell, who was an eloquent advocate of the proposition that intelligent beings inhabit the planet Mars. And yet, through all of this, while his imagination was intensely stimulated, Goddard managed to retain a sense of skepticism very rare in young people given to interplanetary epiphanies high up in cherry trees: "The actual conditions may be entirely different . . . from those which Professor Pickering suggests . . . The only antidote for fallacies is—in a word—to take nothing for granted."

On January 2, 1902, we know from Goddard's notebook, he wrote an essay on "The Habitability of Other Worlds." The paper had not been found among Goddard's writings, which seemed to me a great pity, since it might have given us a better understanding of the extent to which the search for extraterrestrial life was a prime motive in Goddard's lifework.*

* In a commencement address at Clark University on May 18, 1978, I made some similar remarks. Dorothy Mosakowski in the Rare Book Room at Clark's Goddard Memorial Library then searched for and found this little essay which had been listed as lost. In it we discover that Goddard was attracted to but cautious about the possibility of life on Mars, certain of

In his early postdoctoral years Goddard successfully pursued an experimental verification of his ideas on solid- and liquid-fueled rocket flight. In this endeavor he was supported principally by two men: Charles Greeley Abbott and George Ellery Hale. Abbott was then a young scientist at the Smithsonian Institution, of which he later became secretary, the quaint designation by which the executive officer of that organization is still known. Hale was the driving force behind American observational astronomy at the time; before he died he had founded the Yerkes, Mount Wilson and Mount Palomar observatories, each housing, in its time, the largest telescope in the world.

Both Abbott and Hale were solar physicists, and it seems clear that both had been captured by the young Goddard's vision of a rocket sailing free above the obscuring blanket of the Earth's atmosphere, able to view the Sun and stars unimpeded. But Goddard soared far beyond this daring vision. He talked and wrote of experiments on the composition and circulation of the upper atmosphere of the Earth, of performing gamma-ray and ultraviolet observations of the Sun and stars from above the Earth's atmosphere. He conceived of a space vehicle passing 1,000 miles above the surface of Mars—by a curious historical accident just the low point in the orbits of the Mariner 9 and Viking spacecraft. Goddard calculated that a reasonably sized telescope at such a vantage point would be able to photograph features tens of meters across on the surface of the Red Planet, which is the resolution of the Viking orbiter cameras. He conceived of slow interstellar flight at velocities and time scales just equivalent to that of the Pioneer 10 and 11 spacecraft, our first interstellar emissaries.

---

the existence of extrasolar planetary systems and deduced "that among these countless planets there are conditions of heat and light equivalent to those we experience; and if this is the case, and the planet is near our age and size, there may very likely exist human beings like ourselves, probably with strange costumes and still stranger manners." But he also says: "It is for the distant future to answer if we will ever realize truth from our surmises."

Goddard's spirit soared higher still. He conceived, not casually but quite seriously, of solar-powered spacecraft, and in a time when any practical application of nuclear energy was publicly ridiculed, nuclear propulsion for spacecraft over vast interstellar distances. Goddard imagined a time in the far distant future when the Sun has grown cold and the solar system become uninhabitable, when manned interstellar spacecraft would be outfitted by our remote descendants, to visit the stars—not merely the nearby stars, but also remote star clusters in the Milky Way Galaxy. He could not imagine relativistic spaceflight and so hypothesized a method of suspended animation of the human crew or —even more imaginative—a means of sending the genetic material of human beings which would automatically, at some very distant time, be allowed to recombine and produce a new generation of people.

"With each expedition," he wrote, "there should be taken all the knowledge, literature, art (in a condensed form), and description of tools, appliances, and processes, in as condensed, light, and indestructible a form as possible, so that the new civilization could begin where the old ended." These final speculations, entitled "The Last Migration," were sealed in an envelope with instructions to be read "only by an optimist." And that he surely was—not a Pollyanna who chooses to ignore the problems and evils of our times, but rather, a man committed to the improvement of the human condition and the creation of a vast prospect for the future of our species.

Goddard's dedication to Mars was never far from his mind. In the wake of one of his first experimental successes, he was induced to write a press release on the details of his launch and its ultimate significance. He wished to discuss spacecraft to Mars but was dissuaded on the ground that this was too fantastic. As a compromise he talked about sending a quantity of magnesium flash powder which would make a visible bright flare on the Moon when it landed. This caused a sensation in the press. Goddard was for many years after disparagingly referred to as "the Moon Man," and he remained rueful about his relations with the press ever after. (An editorial in the New York *Times* which criticized Goddard for having "forgotten" that a rocket will not work in the vacuum of space because it has nothing to push against may have contributed to his unease. The *Times* discovered Newton's third law of motion and retracted its error only in the age of Apollo.) Goddard mused: "From that day, the whole thing was summed up, in the public mind, in the words 'moon rocket'; and thus it happened that in trying to minimize the sensational side, I had really made more of a stir than if I had discussed transportation to Mars, which would probably have been considered as ridiculous by the press representative and doubtless never mentioned."

Goddard's notebooks are not filled with psychological insights. That was not, at least not very much, the spirit of the times in which he lived.* But there is a remark in Goddard's notebooks that can be only a flash of poignant self-insight: "God pity a one-dream man." That surely is what Goddard was. He knew great satisfaction in seeing the advances in rocket technology, but it must have been agonizingly slow. There are so many letters from Abbott urging faster progress, and so many responses from Goddard citing practical impediments. Goddard never lived to see the beginning of rocket astronomy and high-altitude meteorology, much less flights to the Moon or planets.

But all these things are happening because of what are very clearly the technological fruits of Goddard's genius. October 19, 1976, was the 77th Anniversary Day of the Martian vision of Robert H. Goddard. On that day there were two functioning orbiters and two

* Although, remarkably, he was in Worcester in the year 1909 when Sigmund Freud and Carl Gustav Jung gave the first comprehensive discussion in the English language of those institutionalized insights called psychoanalysis. Many American psychiatrists got their first glimpses of the subject from Freud's Clark University lectures. One wonders if the middle-aged bearded Viennese physician and the young mustachioed American physics graduate student nodded to each other in passing on the Clark University campus, on their way to their separate destinies.

working landers on Mars, the Viking spacecraft whose origins can be traced with utter confidence back to a boy in a cherry tree in a New England autumn in 1899. Among its many other objectives, Viking had the task of checking out the possibility of life on Mars, the prospect that was so powerful a motivation for Goddard so many years before. Curiously, we are still not sure what the Viking biology results mean. Some think that microbial life may have been discovered; others think it unlikely. It is clear that a major program of future exploration of Mars will be required to understand just where in cosmic evolution this neighboring world lies and what its connection is with the state of evolution on our own planet.

From its earliest stages, rocket technology developed because of an interest in life on other worlds. And now that we have landed on Mars, obtained tantalizing and enigmatic biological results, the follow-on missions—roving vehicles and returned sample canisters—in turn require further developments in spacecraft technology, a mutual causality that I think Goddard would have appreciated.

CHAPTER 19

EXPERIMENTS IN SPACE

◆◆◆

We ever long for visions of beauty,
We ever dream of unknown worlds.

MAXIM GORKY

UNTIL RELATIVELY recently, astronomy suffered from a serious impediment and remarkable peculiarity: it was the only thoroughly nonexperimental science. The materials of study were all up there, and we and our machines were all down here.

No other science was so severely constrained. In physics and chemistry, of course, all is forged on the anvil of experiment, and those who doubt a given conclusion are free to perform a wide range of alternative manipulations of matter and energy in an attempt to extract contradictions or alternative explanations. Evolutionary biologists, even those of very patient temperaments, cannot afford to wait a few million years to observe one species evolve into another. But experiments on common amino acid sequences, enzyme structure, nucleic acid codes, chromosomal banding, and anatomy, physiology and behavior make a compelling case for the fact that evolution has occurred and clearly show which plant or animal groups (such as human beings) are related to which others (such as the great apes).

It is true that geophysicists, studying the deep interior of the Earth, cannot travel to the Wiechert discontinuity between core and mantle, or (just yet) to the Mohorovicic discontinuity between mantle and crust. But batholiths, extruded from the deep interior, can be found here and there on the surface and examined. The geophysicists have relied largely on seismic data, and here, like astronomers, they could not force the favors of nature but were compelled to await their voluntary bestowal—for example, in a seismic event situated on the other side of the Earth so that one of two nearby seismometers would be in the shadow of the Earth's core and the other not. But impatient seismologists can and have set off their own chemical and nuclear explosions to ring Earth like a bell. And there are intriguing recent hints that it may be possible to turn earthquakes on and off. Those geologists intolerant of inferential reasoning could always go to the field and examine contemporary erosion processes. But there was no exact astronomical equivalent of the hard-rock geologist.

We have been restricted to the electromagnetic radiation reflected and emitted by astronomical objects. We have not been able to examine pieces of stars or planets* in our laboratories or to fly into such objects to examine them *in situ*. Ground-based passive observations have restricted us to a narrow fraction of the conceivable data on astronomical objects. Our position has been much worse than that of the fabled six blind men in pursuit of the nature of the elephant. It has been more like one blind man in a zoo. We were standing there for centuries stroking a left hind foot. It is not surprising that we did not deduce tusks, or notice that the foot did not belong to an elephant at all. If, by accident, the orbital plane of the double star was in our line of sight, we would see eclipses; otherwise not. We could not move to a position in space from which the eclipses could be observed. If we were observing a gal-

* With the sole exception of the meteorites (see Chapter 15).

axy when a supernova was exploding, we could examine the supernova spectrum; otherwise not. We do not have the ability to perform experiments on supernova explosions—which is just as well. We could not examine in the laboratory the electrical, thermal, mineralogical and organic chemical properties of the lunar surface. We were restricted to inferences from the visible light reflected and the infrared and radio waves emitted by the moon, aided by occasional natural experiments such as eclipses and lunations.

But all that is gradually changing. Ground-based astronomers have, at least for nearby objects, an experimental tool: radar astronomy. At our convenience, at our choice of frequency, polarization, bandpass and pulse length, we can irradiate a nearby moon or planet with microwaves and examine the returned signal. We can wait for the object to rotate underneath the beam and illuminate some other place on its surface. Radar astronomy has delivered a host of new conclusions on the rotation periods of Venus and Mercury, and related problems in the tidal evolution of the solar system, on the craters of Venus, the fragmented surface of the Moon, the elevations of Mars, and the size and composition of the particles in the rings of Saturn. And radar astronomy is just beginning. We are still restricted to low altitudes, and for the outer solar system, to sun-facing hemispheres. But with the newly resurfaced Arecibo telescope of the National Astronomy and Ionosphere Center in Puerto Rico, we will be able to map the surface of Venus to a resolution of 1 kilometer—better than the best ground-based photographic resolution of the lunar surface—and obtain a host of new information on the asteroids, the Galilean satellites of Jupiter and the rings of Saturn. For the first time we are poking around in cosmic stuff, electromagnetically fingering the solar system.

A much more powerful technique of experimental (as opposed to observational) astronomy is spacecraft exploration. We can now travel into the magnetospheres and atmospheres of the planets. We can land on and rove over their surfaces. We can collect material directly from the interplanetary medium. Our first preliminary steps into space have shown us a wide range of phenomena we never knew existed: the Van Allen trapped-particle belts of the earth; the mass concentrations beneath the circular maria of the moon; the sinuous channels and great volcanoes of Mars; the cratered surfaces of Phobos and Deimos. But what I am most struck by is that, before the advent of space vehicles, astronomers did very well—hamstrung though they were. The interpretations of the observations available to them were remarkably good. Space vehicles are ways of checking out the conclusions drawn inferentially by astronomers, a method of determining whether astronomical deductions on very distant objects—objects so far away as to be entirely inaccessible by space vehicles in the near future—should be believed.

ONE OF THE EARLIEST major debates in astronomy was on whether the Earth or the Sun was at the center of the solar system. The Ptolemaic and Copernican views explain the apparent motion of the Moon and planets to comparable precision. For the practical problem of predicting the positions of the Moon and planets as seen from the surface of the Earth, it hardly mattered which hypothesis was adopted. But the philosophical implications of the geocentric and heliocentric hypotheses were quite different. And there were ways of finding out which was right. In the Copernican view, Venus and Mercury should go through phases like the Moon. In the Ptolemaic view, they should not. When Galileo, using one of the first astronomical telescopes, observed a crescent Venus, he knew he had vindicated the Copernican hypothesis.

But space vehicles provide a more immediate test. According to Ptolemy, the planets are affixed to immense crystalline spheres. But when Mariner 2 or Pioneer 10 penetrated the locales of Ptolemy's supposed crystal spheres, no impediment to their motion was detected; and, more directly, the acoustic and other micrometeorite detectors heard not even the faintest whisper of tinkling, much less the sound of smashed crystal.

There is something very satisfying and immediate about this sort of test. There are probably no Ptolemaists in our midst. But there might be some with lingering doubts about whether Venus could not be made to go through phases in some modified geocentric hypothesis. Those people can now rest easy.

Before space vehicles, the German astrophysicist Ludwig Biermann was intrigued by the observations of the apparent acceleration of bright knots in the well-developed tails of comets passing through the inner solar system. Biermann showed that the radiation pressure of sunlight was inadequate to account for the observed acceleration and made the novel suggestion that there were charged particles streaming out from the Sun which, in interaction with the comet, produced the observed acceleration. Well, maybe. But could it not be equally due to, say, chemical explosions in the nucleus of the comet? Or some other explanation? But the first successful interplanetary spacecraft, Mariner 2, in the course of its fly-by of Venus, determined the existence of a solar wind with velocities and electron densities in just the range that Biermann had calculated would be necessary to accelerate his knots.

In the same period there was a debate on the nature of the solar wind. In one view, that of Eugene Parker of the University of Chicago, it was caused by hydrodynamical flow out from the Sun; in another view, by evaporation from the top of the solar atmosphere. In the hydrodynamic explanation there should be no fractionation by mass; that is, the atomic composition of the solar wind should be the same as that of the Sun. But in the evaporation hypothesis, the lighter atoms escape the Sun's gravity more easily, and heavy elements should be preferentially depleted in the solar wind. Interplanetary spacecraft have found that the ratio of hydrogen to helium in the solar wind is precisely that in the Sun, and have thereby provided convincing support for the hydrodynamic hypothesis of the origin of the solar wind.

In these examples from solar wind physics, we find that the spacecraft experiments provided the means for making critical judgments among competing hypotheses. In retrospect, we find that there were astronomers such as Biermann and Parker who were right for the right reasons. But there were others, equally bright, who disbelieved them and might have gone on disbelieving them had not the critical spacecraft experiments been performed. What is remarkable is not that there were alternative hypotheses which we now see to be incorrect, but rather that on the basis of the very meager data available *anyone* was smart enough to divine the correct answer—inferentially, using intuition, physics and common sense.

Before the Apollo missions, the uppermost layer of the lunar surface could be examined by visible, infrared and radio observations during both lunations and eclipses, and the polarization of sunlight reflected off the lunar surface had been measured. From these observations, Thomas Gold of Cornell University prepared a dark powder which, in the laboratory, reproduced the observed properties of the lunar surface very well. This "Golddust" can even be purchased for a modest price from the Edmund Scientific Company. A naked-eye comparison of lunar dust returned by Apollo astronauts with Golddust shows them to be almost indistinguishable. In particle-size distribution, and in electrical and thermal properties, they are a very close match. However, their chemical compositions are very different. Golddust is primarily Portland cement, charcoal and hairspray. The moon has a less exotic composition. But the observed lunar properties available to Gold before Apollo did not strongly depend on the chemical composition of the lunar surface. He was able to deduce very well that fraction of the lunar-surface properties which was relevant to pre-1969 observations of the Moon.

From the study of the available radio and radar data, we were able to deduce the high surface temperature and high surface pressures of Venus before the first Soviet Venera entry probe made *in situ* observations on the atmosphere, and subsequent Venera probes on the surface. Likewise, we correctly deduced the existence of elevation differences on Mars as great as 20 kilometers, although we were mistaken in thinking that dark areas were systematically at high elevations on the planet.*

Perhaps one of the most interesting such confrontations of astronomical inference with spacecraft observations is the case of the magnetosphere of Jupiter. In 1955 Kenneth Franklin and Bernard Burke were testing a radio telescope near Washington, D.C., intended for mapping galactic radio emission at a frequency of 22 Hertz. They noticed a regularly recurring interference on their records, which they at first thought was due to some conventional source of radio noise—such as a faulty ignition system on some nearby tractor. But they soon discovered that the timing of the interference corresponded perfectly well with transits overhead of the planet Jupiter. They had discovered that Jupiter was a powerful source of decameter radio emission.

Subsequently Jupiter was found to be a bright source at decimeter wavelengths as well. But the spectrum was very peculiar. At a wavelength of a few centimeters, very low temperatures of around 140°K were found—temperatures comparable to those uncovered for Jupiter at infrared wavelengths. But at decimeter wavelengths—up to one meter—the brightness temperature increased very rapidly with wavelength, approaching 100,000°K. This was too high a temperature for thermal emission—the radio radiation that all objects put out, simply because they are at a temperature above absolute zero.

Frank Drake, then of the National Radio Astronomy Observatory, proposed in 1959 that this spectrum implied that Jupiter was a source of synchrotron emission—the radiation that charged particles emit in their direction of motion when traveling close to the speed of light. On Earth, synchrotrons are convenient devices used in nuclear physics for accelerating electrons and protons to such high velocities, and it is in synchrotrons that

* I have discussed these successful inferences and their spacecraft confirmations in Chapters 12, 16 and 17 of *The Cosmic Connection.*

such emission was first generally studied. Synchrotron emission is polarized, and the fact that the decimeter radiation from Jupiter is also polarized was an additional point in favor of Drake's hypothesis. Drake suggested that Jupiter was surrounded by a vast belt of relativistic charged particles similar to the Van Allen radiation belt around the Earth, which had then just been discovered. If so, the decimeter emitting region should be much larger than the optical size of Jupiter. But conventional radio telescopes have inadequate angular resolution to make out any spatial detail whatever at the range of Jupiter. A radio interferometer can achieve such resolution, however. In the spring of 1960, very soon after the suggestion was made, V. Radhakrishnan and his colleagues at the California Institute of Technology employed an interferometer composed of two 90-foot-diameter antennas mounted on railroad tracks and separable by almost a third of a mile. They found that the region of decimeter emission around Jupiter was considerably larger than the ordinary optical disc of Jupiter, confirming Drake's proposal.

Subsequent higher-resolution radio interferometry has shown Jupiter to be flanked by two symmetric "ears" of radio-wave emission with the same general configuration as the Van Allen radiation belts of the Earth. The general picture has evolved that on both planets electrons and protons from the solar wind are trapped and accelerated by the planetary magnetic dipole field and are constrained to spiral along the planet's lines of magnetic force, bouncing from one magnetic pole to the other. The radio-emitting region around Jupiter is identified with its magnetosphere. The stronger the magnetic field, the farther out from the planet the boundary of the magnetic field will be. In addition, matching the observed radio spectrum from synchrotron emission theory specifies a magnetic field strength. The field strength could not be specified to very great precision but most estimates from radio astronomy in the late 1960s and early 1970s were in the range of 5 to 30 gauss, some ten to sixty times the surface magnetic field of the Earth at the equator.

Radhakrishnan and colleagues also found that the polarization of the decimeter waves from Jupiter varied regularly as the planet rotated, as if the Jovian radiation belts were wobbling with respect to the line of sight. They proposed that this was due to a 9-degree tilt between the axis of rotation and the magnetic axis of the planet—not very different from the displacement between the north geographic and the north magnetic poles of Earth. Subsequent studies of the decimeter and decameter emission by James Warwick of the University of Colorado and others suggested that the magnetic axis of Jupiter is displaced a small fraction of a Jupiter radius from the axis of rotation, quite different from the terrestrial case, where both axes intersect at the center of the Earth. It was also concluded that the south magnetic pole of Jupiter was in the northern hemisphere; that is, that a north-seeking compass on Jupiter would point south. There is nothing very bizarre about this suggestion. The Earth's magnetic field has flipped its direction many times during its history, and it is only by definition that the north magnetic pole is in the northern hemisphere of the Earth at the present time. From the intensity of the decimeter and decameter emission, astronomers also calculated what the energies and fluxes of electrons and protons in the Jovian magnetosphere might be.

This is a very rich array of conclusions. But all of it is remarkably inferéntial. The whole elaborate superstructure was put to a critical test on December 3, 1973, when the Pioneer 10 spacecraft flew through the Jovian magnetosphere. There were magnetometers aboard, which measured the strength and direction of the magnetic field at various positions in the magnetosphere; and there was a variety of charged-particle detectors, which measured energies and fluxes of the trapped electrons and protons. It is a stunning fact that virtually every one of the radio astronomical inferences was roughly confirmed by Pioneer 10 and its successor spacecraft, Pioneer 11. The surface equatorial magnetic field on Jupiter is about 6 gauss and larger at the poles. The inclination of the magnetic to the rotational axis is about 10 degrees. The magnetic axis can be described as apparently displaced about one quarter of a Jovian radius from the center of the planet. Farther out than three Jupiter radii, the magnetic field is approximately that of a dipole; closer in, it is much more complex than had been estimated.

The flux of charged particles received by Pioneer 10 along its trajectory through the magnetosphere was considerably larger than had been anticipated—but not so large as to inactivate the spacecraft. The survival of Pioneer 10 and 11 through the Jovian magnetosphere was more the result of good luck and good engineering than of the accuracy of pre-Pioneer magnetospheric theories.

In general, the synchrotron theory of the decimeter emission from Jupiter is confirmed. All those radio astronomers turn out to have known what they were doing. We can now believe, with much greater confidence than heretofore, deductions made from synchrotron physics and applied to other, more distant and less accessible comic objects, such as pulsars, quasars or supernova remnants. In fact, the theories can now be recalibrated and their accuracy improved. Theoretical radio astronomy has for the first time been put to a critical experimental test—and it has passed with flying colors. Of the many major findings by Pioneer 10 and 11, I think this is its greatest triumph: it has confirmed our understanding of an important branch of cosmic physics.

There is much about the Jovian magnetosphere and radio emissions that we still do not understand. The details of the decameter emissions are still deeply mysterious. Why are there localized sources of decameter emission on Jupiter probably less than 100 kilometers in size? What are these emission sources? Why do the decameter emission regions rotate about the planet with a very high time precision—better than seven significant figures—but different from the rotation periods of visible features in the Jovian clouds? Why do the decameter bursts have a very intricate (submillisecond) fine structure? Why are the decameter sources beamed—

that is, not emitting in all directions equally? Why are the decameter sources intermittent—that is, not "on" all the time?

These mysterious properties of the Jovian decameter emission are reminiscent of the properties of pulsars. Typical pulsars have magnetic fields a trillion times larger than Jupiter's; they rotate 100,000 times faster; they are a thousandth as old; they are a thousand times more massive. The boundary of the Jovian magnetosphere moves at less than one thousandth of the speed of the light cone of a pulsar. Nevertheless, it is possible that Jupiter is a kind of pulsar that failed, a local and quite unprepossessing model of the rapidly rotating neutron stars, which are one end product of stellar evolution. Major insights into the still baffling problems of pulsar emission mechanisms and magnetosphere geometries may follow from close-up spacecraft observation of Jovian decameter emission—for example, by NASA's Voyager and Galileo missions.

EXPERIMENTAL ASTROPHYSICS is developing rapidly. In another few decades at the very latest, we should see direct experimental investigation of the interstellar medium: the heliopause—the boundary between the region dominated by the solar wind and that dominated by the interstellar plasma—is estimated to lie at not much more than 100 astronomical units (9.3 billion miles) from the Earth. (Now, if there were only a local solar system quasar and a backyard black hole—nothing fancy, you understand, just little baby ones—we might with *in situ* spacecraft measurements check out the greater body of modern astrophysical speculation.)

If we can judge by past experience, each future venture in experimental spacecraft astrophysics will find that (a) a major school of astrophysicists was entirely right; (b) no one agreed on which school it was that was right until the spacecraft results were in; and (c) an entire new corpus of still more fascinating and fundamental problems was unveiled by the space vehicle results.

CHAPTER 20

IN DEFENSE OF ROBOTS

◆◆◆

WILLIAM SHAKESPEARE,
Thou com'st in such a questionable shape
That I will speak to thee . . .
Hamlet, Act I, Scene 4

THE WORD "ROBOT," first introduced by the Czech writer Karel Čapek, is derived from the Slavic root for "worker." But it signifies a machine rather than a human worker. Robots, especially robots in space, have often received derogatory notices in the press. We read that a human being was necessary to make the terminal landing adjustments on Apollo 11, without which the first manned lunar landing would have ended in disaster; that a mobile robot on the Martian surface could never be as clever as astronauts in selecting samples to be returned to Earth-bound geologists; and that machines could never have repaired, as men did, the Skylab sunshade, so vital for the continuance of the Skylab mission.

But all these comparisons turn out, naturally enough, to have been written by humans. I wonder if a small self-congratulatory element, a whiff of human chauvinism, has not crept into these judgments. Just as whites can sometimes detect racism and men can occasionally discern sexism, I wonder whether we cannot here glimpse some comparable affliction of the human spirit —a disease that as yet has no name. The word "anthropocentrism" does not mean quite the same thing. The word "humanism" has been pre-empted by other and more benign activities of our kind. From the analogy with sexism and racism I suppose the name for this malady is "speciesism"—the prejudice that there are no beings so fine, so capable, so reliable as human beings.

This is a prejudice because it is, at the very least, a prejudgment, a conclusion drawn before all the facts are in. Such comparisons of men and machines in space are comparisons of smart men and dumb machines. We have not asked what sorts of machines could have been built for the $30-or-so billion that the Apollo and Skylab missions cost.

Each human being is a superbly constructed, astonishingly compact, self-ambulatory computer—capable on occasion of independent decision making and real control of his or her environment. And, as the old joke goes, these computers can be constructed by unskilled labor. But there are serious limitations to employing human beings in certain environments. Without a great deal of protection, human beings would be inconvenienced on the ocean floor, the surface of Venus, the deep interior of Jupiter, or even on long space missions. Perhaps the only interesting results of Skylab that could not have been obtained by machines is that human beings in space for a period of months undergo a serious loss of bone calcium and phosphorus—which seems to imply that human beings may be incapacitated under 0 g for missions of six to nine months or longer. But the minimum interplanetary voyages have characteristic times of a year or two. Because we value human beings highly, we are reluctant to send them on very risky missions. If we do send human beings to exotic environments, we must also send along their food, their air, their water, amenities for entertainment and waste recycling, and companions. By comparison, machines require no elaborate life-support systems, no entertainment, no companionship, and we do not yet feel any strong ethical prohibitions against sending machines on one-way, or suicide, missions.

Certainly, for simple missions, machines have proved themselves many times over. Unmanned vehicles have performed the first photography of the whole Earth and of the far side of the Moon; the first landings on the Moon, Mars and Venus; and the first thorough orbital reconnaissance of another planet, in the Mariner 9 and Viking missions to Mars. Here on Earth it is increasingly common for high-technology manufacturing—for example, chemical and pharmaceutical plants—to be performed largely or entirely under computer control. In all these activities machines are able, to some extent, to sense errors, to correct mistakes, to alert human controllers some great distance away about perceived problems.

The powerful abilities of computing machines to do arithmetic—hundreds of millions of times faster than unaided human beings—are legendary. But what about really difficult matters? Can machines in any sense think through a new problem? Can they make discussions of the branched-contingency tree variety which we think of as characteristically human? (That is, I ask Question 1; if the answer is A, I ask Question 2; but if the answer is B, I ask Question 3; and so on.) Some decades ago the English mathematician A. M. Turing described what would be necessary for him to believe in machine intelligence. The condition was simply that he could be in teletype communication with a machine and be unable to tell that it was not a human being. Turing imagined a conversation between a man and a machine of the following quality:

INTERROGATOR: In the first line of your sonnet which reads "Shall I compare thee to a Summer's day," would not "a Spring day" do as well or better?

WITNESS: It wouldn't scan.

INTERROGATOR: How about "a Winter's day"? That would scan all right.

WITNESS: Yes, but nobody wants to be compared to a Winter's day.

INTERROGATOR: Would you say Mr. Pickwick reminded you of Christmas?

WITNESS: In a way.

INTERROGATOR: Yet Christmas is a Winter's day, and I do not think Mr. Pickwick would mind the comparison.

WITNESS: I don't think you're serious. By a Winter's day one means a typical Winter's day, rather than a special one like Christmas.

No device of this sophistication has yet been built, although I am not sure how many humans would pass Turing's human test. But the amount of effort and money put into artificial intelligence has been quite limited, and there are only about a half-dozen major centers of such activity in the world. One of the more striking results obtained in a very limited universe of discourse—that of children's blocks—has come from the work of Terry Winograd, then at the Massachusetts Institute of Technology. Here is a dialogue between man and machine, the machine having been programmed by Winograd:

Q: What does the box contain?
A: The blue pyramid and the blue box.
Q: How many blocks are not in the box?
A: Four of them.
Q: Is at least one of them narrower than the one which I told you to pick up?
A: Yes, the red cube.
Q: Is there a steeple?
A: Sorry, I don't know the word "steeple."
Q: A steeple is a stack which contains two green cubes and a pyramid, for example.
A: I understand.
Q: Are there any steeples now?
A: No.
Q: Build one.

The machine does. It reaches out its mechanical hand, moves the cubes and builds the structure that has just been described.

Q: When did you pick up the pyramid?
A: While I was stacking up the red cube, the large red block and the large green cube.
Q: Why?
A: To get rid of it.
Q: Why did you clear off that cube?
A: To put it on the large green cube.

As another example, a machine psychiatrist has been developed by Joseph Weizenbaum, also at MIT. This is a much simpler program than Winograd's, since it involves nondirective psychotherapy, which, generally speaking, requires extremely simple-minded computer programs. Here is a striking dialogue between a human being and a computer psychotherapist. (There is certainly a selection effect here. Many other such machine/human psychiatric dialogues are not nearly so striking. But I suppose most human/human psychiatric dialogues are not very striking either.) In this interaction, in order to spell out clearly the respective roles of therapist and patient, the computer therapist types upper case while the patient types lower case:

men are all alike
IN WHAT WAY
they're always bugging us about something or other
CAN YOU THINK OF A SPECIFIC EXAMPLE
well, my boy friend made me come here
YOUR BOY FRIEND MADE YOU COME HERE
he noticed that i'm depressed much of the time
I AM SORRY TO HEAR YOU ARE DEPRESSED
he's right. i am unhappy
DO YOU THINK COMING HERE WILL HELP YOU NOT TO BE UNHAPPY
i need some help, that much is clear
WHAT WOULD IT MEAN TO YOU IF YOU GOT SOME HELP

perhaps i could learn to get along with my mother
TELL ME MORE ABOUT YOUR FAMILY
my mother takes care of me
WHO ELSE IN YOUR FAMILY TAKES CARE OF YOU
my father
YOUR FATHER
you are like my father in some ways
WHAT RESEMBLANCE DO YOU SEE
you are not very aggressive, but i don't think you want me to notice that
WHAT MAKES YOU THINK I AM NOT VERY AGGRESSIVE
you don't argue with me
WHY DO YOU THINK I DON'T ARGUE WITH YOU
you are afraid of me
DOES IT PLEASE YOU TO BELIEVE I AM AFRAID OF YOU
it pleases my father to think i am afraid of him
WHAT ELSE COMES TO MIND WHEN YOU THINK OF YOUR FATHER
bullies
DOES THAT HAVE ANYTHING TO DO WITH THE FACT THAT YOUR BOY FRIEND MADE YOU COME HERE

This astonishing—one is very tempted to say "perceptive"—response from the computer is, of course, preprogrammed. But, then, so are the responses of human psychotherapists. In a time when more and more people in our society seem to be in need of psychiatric counseling, and when time-sharing of computers is widespread, I can even imagine the development of a network of computer psychotherapeutic terminals, something like arrays of large telephone booths, in which, for a few dollars a session, we are able to talk to an attentive, tested and largely nondirective psychotherapist. Ensuring the confidentiality of the psychiatric dialogue is one of several important steps still to be worked out.

ANOTHER SIGN of the intellectual accomplishments of machines is in games. Even exceptionally simple computers—those that can be wired by a bright ten-year-old—can be programmed to play perfect tic-tac-toe.

Some computers can play world-class checkers. Chess is of course a much more complicated game than tic-tac-toe or checkers. Here programming a machine to win is more difficult, and novel strategies have been used, including several rather successful attempts to have a computer learn from its own experience in playing previous chess games. Computers can learn, for example, empirically the rule that it is better in the beginning game to control the center of the chessboard than the periphery. The ten best chess players in the world still have nothing to fear from any present computer. But the situation is changing. Recently a computer for the first time did well enough to enter the Minnesota State Chess Open. This may be the first time that a non-human has entered a major sporting event on the planet Earth (and I cannot help but wonder if robot golfers and designated hitters may be attempted sometime in the next decade, to say nothing of dolphins in free-style competition). The computer did not win the Chess Open, but this is the first time one has done well enough to enter such a competition. Chess-playing computers are improving extremely rapidly.

I have heard machines demeaned (often with a just audible sigh of relief) for the fact that chess is an area where human beings are still superior. This reminds me very much of the old joke in which a stranger remarks with wonder on the accomplishments of a checker-playing dog. The dog's owner replies, "Oh, it's not all that remarkable. He loses two games out of three." A machine that plays chess in the middle range of human expertise is a very capable machine; even if there are thousands of better human chess players, there are millions who are worse. To play chess requires strategy, foresight, analytical powers, and the ability to cross-correlate large numbers of variables and to learn from experience. These are excellent qualities in those whose job it is to discover and explore, as well as those who watch the baby and walk the dog.

With this as a more or less representative set of examples of the state of development of machine intelligence, I think it is clear that a major effort over the next decade could produce much more sophisticated examples. This is also the opinion of most of the workers in machine intelligence.

In thinking about this next generation of machine intelligence, it is important to distinguish between self-controlled and remotely controlled robots. A self-controlled robot has its intelligence within it; a remotely controlled robot has its intelligence at some other place, and its successful operation depends upon close communication between its central computer and itself. There are, of course, intermediate cases where the machine may be partly self-activated and partly remotely controlled. It is this mix of remote and *in situ* control that seems to offer the highest efficiency for the near future.

For example, we can imagine a machine designed for the mining of the ocean floor. There are enormous quantities of manganese nodules littering the abyssal depths. They were once thought to have been produced by meteorite infall on Earth, but are now believed to be formed occasionally in vast manganese fountains produced by the internal tectonic activity of the Earth. Many other scarce and industrially valuable minerals are likewise to be found on the deep ocean bottom. We have the capability today to design devices that systematically swim over or crawl upon the ocean floor; that are able to perform spectrometric and other chemical examinations of the surface material; that can automatically radio back to ship or land all findings; and that can mark the locales of especially valuable deposits—for example, by low-frequency radio-homing devices. The radio beacon will then direct great mining machines to the appropriate locales. The present state of the art in deep-sea submersibles and in spacecraft environmental sensors is clearly compatible with the development of such devices. Similar remarks can be made for off-shore oil drilling, for coal and other subterranean mineral mining, and so on. The likely economic returns from such devices would pay not only for their development, but for the entire space program many times over.

When the machines are faced with particularly difficult situations, they can be programmed to recognize that the situations are beyond their abilities and to inquire of human operators—working in safe and pleasant environments—what to do next. The examples just given are of devices that are largely self-controlled. The reverse also is possible, and a great deal of very preliminary work along these lines has been performed in the remote handling of highly radioactive materials in laboratories of the U.S. Department of Energy. Here I imagine a human being who is connected by radio link with a mobile machine. The operator is in Manila, say; the machine in the Mindanao Deep. The operator is attached to an array of electronic relays, which transmits and amplifies his movements to the machine and which can, conversely, carry what the machine finds back to his senses. So when the operator turns his head to the left, the television cameras on the machine turn left, and the operator sees on a great hemispherical television screen around him the scene the machine's searchlights and cameras have revealed. When the operator in Manila takes a few strides forward in his wired suit, the machine in the abyssal depths ambles a few feet forward. When the operator reaches out his hand, the mechanical arm of the machine likewise extends itself; and the precision of the man/machine interaction is such that precise manipulation of material at the ocean bottom by the machine's fingers is possible. With such devices, human beings can enter environments otherwise closed to them forever.

In the exploration of Mars, unmanned vehicles have already soft-landed, and only a little further in the future they will roam about the surface of the Red Planet, as some now do on the Moon. We are not ready for a manned mission to Mars. Some of us are concerned about such missions because of the dangers of carrying terrestrial microbes to Mars, and Martian microbes, if they exist, to Earth, but also because of their enormous expense. The Viking landers deposited on Mars in the summer of 1976 have a very interesting array of sensors and scientific instruments, which are the extension of human senses to an alien environment.

The obvious post-Viking device for Martian exploration, one which takes advantage of the Viking technology, is a Viking Rover in which the equivalent of an entire Viking spacecraft, but with considerably improved science, is put on wheels or tractor treads and permitted to rove slowly over the Martian landscape. But now we come to a new problem, one that is never encountered in machine operation on the Earth's surface. Although Mars is the second closest planet, it is so far from the Earth that the light travel time becomes significant. At a typical relative position of Mars and the Earth, the planet is 20 light-minutes away. Thus, if the spacecraft were confronted with a steep incline, it might send a message of inquiry back to Earth. Forty minutes later the response would arrive saying something like "For heaven's sake, stand dead still." But by then, of course, an unsophisticated machine would have tumbled into the gully. Consequently, any Martian Rover requires slope and roughness sensors. Fortunately, these are readily available and are even seen in some children's toys. When confronted with a precipitous slope or large boulder, the spacecraft would either stop until receiving instructions from the Earth in response to its query (and televised picture of the terrain), or back off and start in another and safer direction.

Much more elaborate contingency decision networks can be built into the onboard computers of spacecraft of the 1980s. For more remote objectives, to be explored further in the future, we can imagine human controllers in orbit around the target planet, or on one of its moons. In the exploration of Jupiter, for example, I can imagine the operators on a small moon outside the fierce Jovian radiation belts, controlling with only a few seconds' delay the responses of a spacecraft floating in the dense Jovian clouds.

Human beings on Earth can also be in such an interaction loop, if they are willing to spend some time on the enterprise. If every decision in Martian exploration must be fed through a human controller on Earth, the Rover can traverse only a few feet an hour. But the lifetimes of such Rovers are so long that a few feet an hour represents a perfectly respectable rate of progress. However, as we imagine expeditions into the farthest reaches of the solar system—and ultimately to the stars—it is clear that self-controlled machine intelligence will assume heavier burdens of responsibility.

In the development of such machines we find a kind of convergent evolution. Viking is, in a curious sense, like some great outsized, clumsily constructed insect. It is not yet ambulatory, and it is certainly incapable of self-reproduction. But it has an exoskeleton, it has a wide range of insectlike sensory organs, and it is about as intelligent as a dragonfly. But Viking has an advantage that insects do not: it can, on occasion, by inquiring of its controllers on Earth, assume the intelligence of a human being—the controllers are able to reprogram the Viking computer on the basis of decisions they make.

As the field of machine intelligence advances and as increasingly distant objects in the solar system become accessible to exploration, we will see the development of increasingly sophisticated onboard computers, slowly climbing the phylogenetic tree from insect intelligence to crocodile intelligence to squirrel intelligence and—in the not very remote future, I think—to dog intelligence. Any flight to the outer solar system must have a computer capable of determining whether it is working properly. There is no possibility of sending to the Earth for a repairman. The machine must be able to sense when it is sick and skillfully doctor its own illnesses. A computer is needed that is able either to fix or replace failed computer, sensor or structural components. Such a computer, which has been called STAR (self-testing and repairing computer), is on the threshold of development. It employs redundant components, as biology does—we have two lungs and two kidneys partly because each is protection against failure of the other. But a computer can be much more redundant than a human being, who has, for example, but one head and one heart.

Because of the weight premium on deep space exploratory ventures, there will be strong pressures for continued miniaturization of intelligent machines. It is clear that remarkable miniaturization has already occurred: vacuum tubes have been replaced by transistors, wired circuits by printed circuit boards, and entire computer systems by silicon-chip microcircuitry. Today a circuit that used to occupy much of a 1930 radio set can be printed on the tip of a pin. If intelligent machines for terrestrial mining and space exploratory applications are pursued, the time cannot be far off when household and other domestic robots will become commercially feasible. Unlike the classical anthropoid robots of science fiction, there is no reason for such machines to look any more human than a vacuum cleaner does. They will be specialized for their functions. But there are many common tasks, ranging from bartending to floor washing, that involve a very limited array of intellectual capabilities, albeit substantial stamina and patience. All-purpose ambulatory household robots, which perform domestic functions as well as a proper nineteenth-century English butler, are probably many decades off. But more specialized machines, each adapted to a specific household function, are probably already on the horizon.

It is possible to imagine many other civic tasks and essential functions of everyday life carried out by intelligent machines. By the early 1970s, garbage collectors in Anchorage, Alaska, and other cities won wage settlements guaranteeing them salaries of about $20,000 per annum. It is possible that the economic pressures alone may make a persuasive case for the development of automated garbage-collecting machines. For the development of domestic and civic robots to be a general civic good, the effective re-employment of those human beings displaced by the robots must, of course, be arranged; but over a human generation that should not be too difficult—particularly if there are enlightened educational reforms. Human beings enjoy learning.

We appear to be on the verge of developing a wide variety of intelligent machines capable of performing tasks too dangerous, too expensive, too onerous or too boring for human beings. The development of such machines is, in my mind, one of the few legitimate "spin-offs" of the space program. The efficient exploitation of energy in agriculture—upon which our survival as a species depends—may even be contingent on the development of such machines. The main obstacle seems to be a very human problem, the quiet feeling that comes stealthily and unbidden, and argues that there is something threatening or "inhuman" about machines performing certain tasks as well as or better than human beings; or a sense of loathing for creatures made of silicon and germanium rather than proteins and nucleic acids. But in many respects our survival as a species depends on our transcending such primitive chauvinisms. In part, our adjustment to intelligent machines is a matter of acclimatization. There are already cardiac pacemakers that can sense the beat of the human heart; only when there is the slightest hint of fibrillation does the pacemaker stimulate the heart. This is a mild but very useful sort of machine intelligence. I cannot imagine the wearer of this device resenting its intelligence. I think in a relatively short period of time there will be a very similar sort of acceptance for much more intelligent and sophisticated machines. There is nothing inhuman about an intelligent machine; it is indeed an expression of those superb intellectual capabilities that only human beings, of all the creatures on our planet, now possess.

CHAPTER 21

THE PAST AND FUTURE OF AMERICAN ASTRONOMY

◆◆◆

> What has been done is little—scarcely a beginning; yet it is much in comparison with the total blank of a century past. And our knowledge will, we are easily persuaded, appear in turn the merest ignorance to those who come after us. Yet it is not to be despised, since by it we reach up groping to touch the hem of the garment of the Most High.
>
> AGNES M. CLERKE,
> *A Popular History of Astronomy*
> (London, Adam and Charles Black, 1893)

THE WORLD has changed since 1899, but there are few fields which have changed more—in the development of fundamental insights and in the discovery of new phenomena—than astronomy. Here are a few titles of recent papers published in the scientific magazines *The Astrophysical Journal* and *Icarus:* "G240-72: A New Magnetic White Dwarf with Unusual Polarization," "Relativistic Stellar Stability: Preferred Frame Effects," "Detection of Interstellar Methylamine," "A New List of 52 Degenerate Stars," "The Age of Alpha Centauri,"

"Do OB Runaways Have Collapsed Companions?," "Finite Nuclear-size Effects on Neutrino-pair Bremsstrahlung in Neutron Stars," "Gravitational Radiation from Stellar Collapse," "A Search for a Cosmological Component of the Soft X-ray Background in the Direction of M31," "The Photochemistry of Hydrocarbons in the Atmosphere of Titan," "The Content of Uranium, Thorium and Potassium in the Rocks of Venus as Measured by Venera 8," "HCN Radio Emission from Comet Kohoutek," "A Radar Brightness and Altitude Image of a Portion of Venus" and "A Mariner 9 Photographic Atlas of the Moons of Mars." Our astronomical ancestors would have extracted a glimmer of meaning from these titles, but I think their principal reaction would have been one of incredulity.

WHEN I WAS ASKED to chair the 75th Anniversary Committee of the American Astronomical Society in 1974, I thought it would provide a pleasant opportunity to acquaint myself with the state of our subject at the end of the past century. I was interested to see where we had been, where we are today, and if possible, something of where we may be going. In 1897 the Yerkes Observatory, then the largest telescope in the world, was given a formal dedication, and a scientific meeting of astronomers and astrophysicists was held in connection with the ceremony. A second meeting was held at the Harvard College Observatory in 1898 and a third at the Yerkes Observatory in 1899, by which time what is now the American Astronomical Society had been officially founded.

The astronomy of 1897 to 1899 seems to have been vigorous, combative, dominated by a few strong personalities and aided by remarkably short publication times. The average time between submission and publication for papers in the *Astrophysical Journal (Ap. J.)* in this period seems to be better than in *Astrophysical Journal Letters* today. The fact that a great many papers were from the Yerkes Observatory, where the journal was edited, may have had something to do with this. The opening of the Yerkes Observatory at Williams Bay, Wisconsin—which has the year 1895 imprinted upon it—was delayed more than a year because of the collapse of the floor, which narrowly missed killing the astronomer E. E. Barnard. The accident is mentioned in *Ap. J.* (6:149), but one finds no hint of negligence there. However, the British journal *Observatory* (20:393), clearly implies careless construction and a cover-up to shield those responsible. We also discover on the same page of *Observatory* that the dedication ceremonies were postponed for some weeks to accommodate the travel schedule of Mr. Yerkes, the robber-baron donor. The *Astrophysical Journal* says that "the dedication ceremonies were necessarily postponed from October 1, 1897," but does not say why.

Ap. J. was edited by George Ellery Hale, the director of the Yerkes Observatory, and by James E. Keeler, who in 1898 became the director of the Lick Observatory on Mount Hamilton in California. However, there was a certain domination of *Ap. J.* by Williams Bay, perhaps because the Lick Observatory dominated the *Publications* of the Astronomical Society of the Pacific *(PASP)* in the same period. Volume 5 of the *Astrophysical Journal* has no fewer than thirteen plates of the Yerkes Observatory, including one of the powerhouse. The first fifty pages of Volume 6 have a dozen more plates of the Yerkes Observatory. The Eastern dominance of the American Astronomical Society is also reflected by the fact that the first president of the Astronomical and Astrophysical Society of America was Simon Newcomb, of the Naval Observatory in Washington, and the first vice presidents, Young and Hale. West Coast astronomers complained about the difficulties in traveling to the third conference of astronomers and astrophysicists at Yerkes and seem to have voiced some pleasure that promised demonstrations with the Yerkes 40-inch refractor for this ceremony had to be postponed because of cloudy weather. This was about the most in the way of interobservatory rancor that can be found in either journal.

But in the same period *Observatory* had a keen nose for American astronomical gossip. From *Observatory* we find that there was a "civil war" at the Lick Observatory and a "scandal" associated with Edward Holden (the director before Keeler), who is said to have permitted rats in the drinking water at Mount Hamilton. It also published a story about a test chemical explosion scheduled to go off in the San Francisco Bay Area and to be monitored by a seismic device on Mount Hamilton. At the appointed moment, no staff member could see any sign of needle deflection except for Holden, who promptly dispatched a messenger down the mountain to alert the world to the great sensitivity of the Lick seismometer. But soon up the mountain came another messenger with the news that the test had been postponed. A much faster messenger was then dispatched to overtake the first and an embarrassment to the Lick Observatory was, *Observatory* notes, narrowly averted.

The youth of American astronomy in this period is eloquently reflected in the proud announcement in 1900 that the Berkeley Astronomical Department would henceforth be independent of the Civil Engineering Department at the University of California. A survey by Professor George Airy, later the British Astronomer Royal, regretted being unable to report on astronomy in America in 1832 because essentially there was none. He would not have said that in 1899.

There is never much sign in these journals of the intrusion of external (as opposed to academic) politics, except for an occasional notice such as the appointment by President McKinley of T. J. J. See as professor of mathematics to the U.S. Navy, and a certain continuing chilliness in scientific debates between the personnel of the Lick and Potsdam (Germany) Observatories.

Some signs of the prevailing attitudes of the 1890s occasionally trickle through. For example, in a description of an eclipse expedition to Siloam, Georgia, on May 28, 1900: "Even some of the whites were lacking in a very deep knowledge of things 'eclipse-wise.' Many thought it was a money-making scheme and what I intended to charge for admission was a very important question, frequently asked. Another idea was that the eclipse could be seen only from the inside of my observatory . . . Just here I wish to express my appreciation of the high moral tone of the community, for, with a population of only 100, including the immediate neighborhood, it sustains 2 white and 2 colored churches and during my stay I did not hear a single profane word . . . As an unsophisticated Yankee in the Southland, unused to Southern ways, I naturally made many little slips that were not considered 'just the thing.' The smiles at my prefixing 'Mr.' to the name of my colored helper caused me to change it to 'Colonel,' which was entirely satisfactory to everybody."

A board of visitors was appointed to resolve some (never publicly specified) problems at the U.S. Naval Observatory. A report of this group—which consisted of two obscure senators and Professors Edward C. Pickering, George C. Comstock and Hale—is illuminating because it mentions dollar amounts. We find that the annual running costs of the major observatories in the world were: Naval Observatory, $85,000; Paris Observatory, $53,000; Greenwich Observatory (England), $49,000; Harvard Observatory, $46,000; and Pulkowa Observatory (Russia), $36,000. The salaries of the two directors of the U.S. Naval Observatory were $4,000 each, and at the Harvard Observatory, $5,000. The distinguished board of visitors recommended that in a "schedule of salaries which could be expected to attract astronomers of the class desired," the salary of directors of observatories should be $6,000. At the Naval Observatory, computers (exclusively human at the time) were paid $1,200 per annum, but at the Harvard Observatory only $500 per annum, and were almost exclusively women. In fact, all salaries at Harvard, except for the director's, were significantly lower than at the Naval Observatory. The committee stated: "The great difference in salaries at Washington and Cambridge, especially for the officers of lower grade, is probably unavoidable. This is partly due to Civil Service Rules." An additional sign of astronomical impecuniosity is the announcement of the post of "volunteer research assistant" at Yerkes, which had no associated pay but which was said to provide good experience for students with higher degrees.

Then, as now, astronomy was besieged by "paradoxers," proponents of fringe or crackpot ideas. One proposed a telescope with ninety-one lenses in series as an alternative to a telescope with a smaller number of lenses of larger aperture. The British in this period were similarly plagued but in perhaps a gentler way. For example, an obituary in the *Monthly Notices* of the Royal Astronomical Society (59:226) of Henry Perigal informs us that the deceased had celebrated his ninety-fourth birthday by becoming a member of the Royal Institution, but was elected a Fellow of the Royal Astronomical Society in 1850. However, "our publications contain nothing from his pen." The obituary describes "the remarkable way in which the charm of Mr. Perigal's personality won him a place which might have seemed impossible of attainment for a man of his views; for there is no masking the fact that he was a paradoxer pure and simple, his main conviction being that the Moon did not rotate, and his main astronomical aim in life being to convince others, and especially young men not hardened in the opposite belief, of their grave error. To this end he made diagrams, constructed models, and wrote poems; bearing with heroic cheerfulness the continued disappointment of finding none of them of any avail. He has, however, done excellent work apart from this unfortunate misunderstanding."

The number of American astronomers in this period was very small. The by-laws of the Astronomical and Astrophysical Society of America state that a quorum is constituted by twenty members. By the year 1900 only nine doctorates had been granted in astronomy in America. In that year there were four astronomical doctorates: two from Columbia University for G. N. Bauer and Carolyn Furness; one from the University of Chicago for Forest Ray Moulton; and one from Princeton University for Henry Norris Russell.

Some idea of what was considered important scientific work in this period can be garnered from the prizes that were awarded. E. E. Barnard received the Gold

Medal of the Royal Astronomical Society in part for his discovery of the Jovian moon Jupiter 5 and for his astronomical photography with a portrait lens. His steamer, however, was caught in an Atlantic storm, and he did not arrive in time for the celebration ceremony. He is described as requiring several days to recover from the storm, whereupon the RAS hospitably gave a second dinner for him. Barnard's lecture seems to have been spectacular and made full use of that recently improved audio-visual aid, the lantern slide projector.

In his discussion of his photograph of the region of the Milky Way near Theta Ophiuchus he concluded that "the entire groundwork of the Milky Way . . . has a substratum of nebulous matter." (Meanwhile H. K. Palmer reported no nebulosity in photographs of the globular cluster M13.) Barnard, who was a superb visual observer, expressed considerable doubts about Percival Lowell's view of an inhabited and canal-infested Mars. In his thanks to Barnard for his lecture, the president of the Royal Astronomical Society, Sir Robert Ball, voiced concern that henceforth he "should regard the canals in Mars with some suspicion, nay, even the seas [of Mars, the dark areas] had partly fallen under a ban. Perhaps the lecturer's recent experiences on the Atlantic might explain something of this mistrust." Lowell's views were not then in favor in England, as another notice in *Observatory* indicated. In response to an inquiry on which books had most pleased and interested him in 1896, Professor Norman Lockyer replied, "*Mars* by Percival Lowell, *Sentimental Tommy* by J. M. Barrie. (No Time for Reading Seriously)."

Prizes in astronomy for 1898 awarded by the Académie Française included one to Seth Chandler for the discovery of the variation in latitude; one to Belopolsky, partly for studies of spectroscopic binary stars; and one to Schott for work on terrestrial magnetism. There was also a prize competition for the best treatise on "the theory of perturbations of Hyperion," a moon of Saturn. We are informed that "the only essay presented was that by Dr. G. W. Hill of Washington to whom the prize was awarded."

The Astronomical Society of the Pacific's Bruce Medal was awarded in 1899 to Dr. Arthur Auwers of Berlin. The dedicatory address included the following remarks: "Today Auwers stands at the head of German astronomy. In him is seen the highest type of investigator in our time, one perhaps better developed in Germany than in any other country. The work of men of this type is marked by minute and careful research, untiring industry in the accumulation of facts, caution in propounding new theories or explanations, and, above all, the absence of effort to gain recognition by being the first to make a discovery." In 1899 the Henry Draper Gold Medal of the National Academy of Sciences was presented for the first time in seven years. The recipient was Keeler. In 1898 Brooks, whose observatory was in Geneva, New York, announced the discovery of his twenty-first comet—which Brooks described as "achieving his majority." Shortly thereafter he received the Lalande Prize of the Académie Française for his record in discovering comets.

In 1897, in connection with an exhibition in Brussels, the Belgian government offered prizes for the solutions of certain problems in astronomy. These problems included the numerical value of the acceleration due to gravity on Earth, the secular acceleration of the Moon, the net motion of the solar system through space, the variation of latitude, the photography of planetary surfaces, and the nature of the canals of Mars. A final topic was the invention of a method to observe the solar corona in the absence of an eclipse. *Monthly Notices* (20:145) commented: ". . . if this pecuniary reward does induce anyone to solve this last problem or in fact any of the others, we think the money will be well spent."

However, reading the scientific papers of this time, one gets the impression that the focus had shifted to other topics than those for which prizes were being given. Sir William and Lady Huggins performed laboratory experiments which showed that at low pressures the emission spectrum of calcium exhibited only the so-called H and K lines. They concluded that the Sun was composed chiefly of hydrogen, helium, "coronium" and calcium. Huggins had earlier established a stellar spectral sequence, which he believed was evolutionary. The Darwinian influence in science was very strong in this period, and among American astronomers T. J. J. See's work was notably dominated by a Darwinian perspective. It is interesting to compare Huggins' spectral sequence with the present Morgan-Keenan spectral types:

HUGGINS' STELLAR SPECTRAL SEQUENCE

Order of Increasing Age	*Star (and modern spectral type in parentheses)*	
Young	Sirius (A1V)	Vega (A0V)
	....	
	Altair (A7 IV–V)	
	Rigel (B8Ia)	
	Deneb (A2Ia)	
	....	
	....	
	Capella (G8, G0)	Sun (G0)
	Arcturus (K1 III)	
	Aldebaran (K5 III)	
Old	Betelgeuse (M2 I)	

Note: The modern stellar spectral sequence runs, from "early" to "late" spectral types, as O, B, A, F, G, K, M. Huggins was very nearly right.

We can see here the origin of the present terms "early" and "late" spectral type, which reflect the Darwinian spirit of late Victorian science. It is also clear that there is a reasonably continuous gradation of spectral types here, and the beginnings—through the later Hertzsprung-Russell diagram—of modern theories of stellar evolution.

There were major developments in physics during this period and readers of *Ap. J.* were alerted to them by the reprinting of summaries of important papers. Experiments were still being performed on the basic radiation laws. In some papers, the level of physical sophistication was not of the highest caliber, as, for example, in an article in *PASP* (11:18) where the linear momentum of Mars is calculated as the single product of the mass of the planet and the linear velocity of the surface. It concluded "the planet, exclusive of the cap, has a momentum of 183 and 3/8 septillion foot pounds." Exponential notation for large numbers was evidently not in wide use.

In this time we have the publication of visual and photographic light curves, for example, of stars in M5; and experiments in filter photography of Orion by Keeler. An obviously exciting topic was time-variable astronomy, which must then have generated something of the excitement that pulsars, quasars and X-ray sources do today. There were many studies of variable velocities in the line of sight from which were derived the orbits of spectroscopic binaries, as well as periodic variations in the apparent velocity of Omicron Ceti from the Doppler displacement of H gamma and other spectral lines.

The first infrared measurements of stars were performed at the Yerkes Observatory by Ernest F. Nichols. The study concludes: "We do not receive from Arcturus more heat than would reach us from a candle at a distance of 5 or 6 miles." No further calculations are given. The first experimental observations of the infrared opacity of carbon dioxide and water vapor were performed in this period by Rubens and Aschkinass, who essentially discovered the ν_2 fundamental of carbon dioxide at 15 microns and the pure rotation spectrum of water.

There is preliminary photographic spectroscopy of the Andromeda nebula by Julius Scheiner at Potsdam, Germany, who concludes correctly that "the previous suspicion that the spiral nebulae are star clusters is now raised to a certainty." As an example of the level of personal vituperation tolerated at this time, the following is an extract from a paper by Scheiner in which W. W. Campbell is criticized: "In the November number of the *Astrophysical Journal,* Professor Campbell attacks, with much indignation, some remarks of mine criticizing his discoveries . . . Such sensitiveness is somewhat surprising on the part of one who is himself given to severely taking others to task. Further, an astronomer who frequently observes phenomena which others cannot see, and fails to see those which others can, must be prepared to have his opinions contested. If, as Professor Campbell complains, I have only supported my views by a single example, I was only withheld by courteous motives from adding another. Namely, the fact that Professor Campbell cannot perceive the lines of aqueous vapor in the spectrum of Mars which were seen by Huggins and Vogel in the first place, and, after Mr. Campbell had called their existence in question, were again seen and identified with certainty by Professor Wilsing and myself." The amount of water vapor in the Martian atmosphere that is now known to exist would have been entirely indetectable by the spectroscopic methods then in use.

Spectroscopy was a dominant element in late-nineteenth-century science. *Ap. J.* was busily publishing Rowland's solar spectrum, which ran to 20,000 wavelengths, each to seven significant figures. It published a major obituary of Bunsen. Occasionally the astronomers took note of the extraordinary nature of their discoveries: "It is simply amazing that the feeble twinkling light of a star can be made to produce such an autographic record of substance and condition of the inconceivably distant luminary." A major topic of debate for the *Astrophysical Journal* was whether spectra should be shown with red to the left or to the right. Those who favored red to the left cited the analogy of the piano (where high frequencies are to the right), but *Ap. J.* opted gamely for red to the right. Some room for compromise was available on whether, in lists of wavelengths, red should be at the top or at the bottom. Feelings ran high, and Huggins wrote to say that "any change . . . would be little less than intolerable." But the *Ap. J.* won anyway.

Another major discussion in this period was on the nature of sunspots. G. Johnstone Stoney proposed that they were caused by a layer of condensed clouds in the photosphere of the Sun. But Wilson and FitzGerald objected to this on the ground that no conceivable condensates could exist at these high temperatures, with the possible exception of carbon. They suggested instead and very vaguely that sunspots are due to "reflection by convection streams of gas." Evershed had a more ingenious idea. He thought that sunspots were holes in the outer photosphere of the Sun, permitting us to see to much greater and hotter depths. But why are they dark? He proposed that all the radiation would be moved from the visible to the inaccessible ultraviolet. This, of course, was before the Planck distribution of radiation from a hot object was understood. It was not at this time thought impossible that the spectral distributions of black bodies of different temperatures should cross; and some experimental curves of this period indeed showed such crossing—due, as we now know, to emissivity and absorptivity differences.

Ramsay had recently discovered the element krypton, which was said to have, among fourteen detectable spectral lines, one at 5570 Å, coincident with "the principal line of the aurora." E. B. Frost concluded: "Thus it seems that the true origin of that hitherto perplexing line has been discovered." We now know it is due to oxygen.

There were a great many papers on instrumental design, one of the more interesting being by Hale. In January 1897 he suggested that both refracting and reflecting telescopes were needed, but noted that there was a clear movement toward reflectors, especially equatorial coude telescopes. In a historical memoir, Hale mentions that the 40-inch lens was available to the Yerkes Observatory only because a previous plan to build a large refractor near Pasadena, California, had fallen through. What, I wonder, would the history of astronomy have been like if the plan had succeeded? Curiously enough, Pasadena seems to have made an offer to the University of Chicago to have the Yerkes Observatory situated there. It would have been a long commute for 1897.

AT THE END of the nineteenth century, solar system studies displayed the same mixture of future promise and current confusion that the stellar work did. One of the most notable papers of the period, by Henry Norris Russell, is called "The Atmosphere of Venus." It is a discussion of the extension of the cusps of the crescent Venus, based in part on the author's observations with the 5-inch *finder* telescope of the "great equatorial" of the Halsted Observatory at Princeton. Perhaps the young Russell was not yet considered fully reliable operating larger telescopes at Princeton. The essence of the analysis is correct by present standards. Russell concluded that refraction of sunlight was not responsible for the extension of the cusps, and that the cause was to be found in the scattering of sunlight: ". . . the atmosphere of Venus, like our own, contains suspended particles of dust or fog of some sort, and . . . what we see is the upper part of this hazy atmosphere, illuminated by rays that have passed close to the planet's surface." He later says that the apparent surface may be a dense cloud layer. The height of the haze is calculated as about 1 kilometer above what we would now call the main cloud deck, a number that is just consistent with limb photography by the Mariner 10 spacecraft. Russell thought, from the work of others, that there was some spectroscopic evidence for water vapor and oxygen in a thin Venus atmosphere. But the essence of his argument has stood the test of time remarkably well.

William H. Pickering's discovery of Phoebe, the outermost satellite of Saturn, was announced; and Andrew E. Douglass of the Lowell Observatory published observations that led him to conclude that Jupiter 3 rotates about one hour slower than its period of revolution, a conclusion incorrect by one hour.

Others who estimated periods of rotation were not quite so successful. For example, there was a Leo Brenner who observed from the Manora Observatory in a place called Lussinpiccolo. Brenner severely criticized Percival Lowell's estimate of the rotation period of Venus. Brenner himself compared two drawings of Venus in white light made by two different people four years apart—from which he deduced a rotation period of 23 hours, 57 minutes and 36.37728 seconds, which he said agreed well with the mean of his own "most reliable" drawings. Considering this, Brenner found it incomprehensible that there could still be partisans of a 224.7-day rotation period and concluded that "an inexperienced observer, an unsuitable telescope, an unhappily chosen eyepiece, a very small diameter of the planet, observed with an insufficient power, and a low declination, all together explained Mr. Lowell's peculiar drawings." The truth, of course, lies not between the extremes of Lowell and Brenner, but rather at the other end of the scale, with a minus sign, a retrograde period of 243 days.

In another communication Herr Brenner begins: "Gentlemen: I have the honor to inform you that Mrs. Manora has discovered a new division in the Saturnian ring system"—from which *we* discover that there is a *Mrs.* Manora at the Manora Observatory in Lussinpiccolo and that she performs observations along with Herr Brenner. Then follows a description of how the Encke, Cassini, Antoniadi, Struve and Manora divisions are all to be kept straight. Only the first two have stood the test of time. Herr Brenner seems to have faded into the mists of the nineteenth century.

AT THE SECOND CONFERENCE of Astronomers and Astrophysicists at Cambridge, there was a paper on the "suggestion" that asteroid rotation, if any, might be deduced from a light curve. But no variation of the brightness with time was found, and Henry Parkhurst concluded: "I think it is safe to dismiss the theory." It is now a cornerstone of asteroid studies.

In a discussion of the thermal properties of the Moon, made independently of the one-dimensional equation of heat conduction but based on laboratory emissivity measurements, Frank Very concluded that a typical lunar daytime temperature is about 100°C—exactly the right answer. His conclusion is worth quoting: "Only the most terrible of Earth's deserts where the burning sands blister the skin, and the men, beasts, and birds drop dead, can approach noontide on the cloudless surface of our satellite. Only the extreme polar latitudes of the

Moon can have an endurable temperature by day, to say nothing of the night, when we should have to become troglodytes to preserve ourselves from such intense cold." The expository styles were often fine.

Earlier in the decade, Maurice Loewy and Pierre Puiseux at the Paris Observatory had published an atlas of lunar photographs, the theoretical consequences of which were discussed in *Ap. J.* (5:51). The Paris group proposed a modified volcanic theory for the origin of the lunar craters, rills and other topographic forms, which was later criticized by E. E. Barnard after he examined the planet with the 40-inch telescope. Barnard was then criticized by the Royal Astronomical Society for his criticism, and so on. One of the arguments in this debate had a deceptive simplicity: volcanoes produce water; there is no water on the moon; therefore the lunar craters are not volcanic. While most of the lunar craters are not volcanic, this is not a convincing argument because it neglects the problem of possible repositories for water. Very's conclusions on the temperature of the lunar poles could have been read with some profit. Water there freezes out as frost. The other possibility is that water might escape from the Moon to space.

This was recognized by Stoney in a remarkable paper called "Of Atmospheres upon Planets and Satellites." He deduced that there should be no lunar atmosphere because of the very rapid escape to space of gases from the low lunar gravity, or any large build-up of the lightest gases, hydrogen and helium, on Earth. He believed that there was no water vapor in the Martian atmosphere and that Mars' atmosphere and caps were probably carbon dioxide. He implied that hydrogen and helium were to be expected on Jupiter, and that Triton, the largest moon of Neptune, might have an atmosphere. Each of these conclusions is in accord with present-day findings or opinions. He also concluded that Titan should be airless, a prediction with which some modern theorists agree—although Titan seems to have another view of the matter (see Chapter 13).

In this period there are also a few breath-taking

speculations, such as one by the Rev. J. M. Bacon that it would be a good idea to perform astronomical observations from high altitudes—from, for example, a free balloon. He suggested that there would be at least two advantages: better seeing and ultraviolet spectroscopy. Goddard later made similar proposals for rocket-launched observatories (Chapter 18).

Hermann Vogel had previously found, by eyeball spectroscopy, an absorption band at 6183 Å in the body of Saturn. Subsequently the International Color Photo Company of Chicago made photographic plates, which were so good that wavelengths as long as H Alpha in the red could be detected for a fifth-magnitude star. This new emulsion was used at Yerkes, and Hale reported that there was no sign of the 6183 Å band for the rings of Saturn. The band is now known to be at 6190 Å and is $6\nu_3$ of methane.

Another reaction to Percival Lowell's writings can be gleaned from the address of James Keeler at the dedication of the Yerkes Observatory:

> It is to be regretted that the habitability of the planets, a subject of which astronomers profess to know little, has been chosen as a theme for exploitation by the romancer, to whom the step from habitability to inhabitants is a very short one. The result of his ingenuity is that fact and fancy become inextricably tangled in the mind of the layman, who learns to regard communication with the inhabitants of Mars as a project deserving serious consideration (for which he may even wish to give money to scientific societies), and who does not know that it is condemned as a vagary by the very men whose labors have excited the imagination of the novelist. When he is made to understand the true state of our knowledge of these subjects, he is much disappointed and feels a certain resentment towards science, as if it had imposed upon him. Science is not responsible for these erroneous ideas, which, having no solid basis, gradually die out and are forgotten.

The address of Simon Newcomb on this occasion contains some remarks which apply generally, if a little idealistically, to the scientific endeavor:

> Is the man thus moved into the exploration of nature by an unconquerable passion more to be envied or pitied? In no other pursuit does such certainty come to him who deserves it. No life is so enjoyable as that whose energies are devoted to following out the inborn impulses of one's nature. The investigator of truth is little subject to the disappointments which await the ambitious man in other fields of activity. It is pleasant to be one of a brotherhood extending over the world in which no rivalry exists except that which comes out of trying to do better work than anyone else, while mutual admiration stifles jealousy . . . As the great captain of industry is moved by the love of wealth and the politician by the love of power, so the astronomer is moved by the love of knowledge for its own sake and not for the sake of its application. Yet he is proud to know that his science has been worth more to mankind than it has cost . . . He feels that man does not live by bread alone. If it is not more than bread to know the place we occupy in the universe, it is certainly something that we should place not far behind the means of subsistence.

AFTER READING through the publications of astronomers three-quarters of a century ago, I felt an irresistible temptation to imagine the 150th Anniversary Meeting of the American Astronomical Society—or whatever name it will have metamorphosed into by then—and guess how our present endeavors will be viewed.

In examining the late-nineteenth-century literature, we are amused at some of the debates on sunspots, and impressed that the Zeeman effect was not considered a laboratory curiosity but something to which astronomers should devote considerable attention. These two threads intertwined, as if prefigured, a few years later in G. E. Hale's discovery of large magnetic field strengths in sunspots.

Likewise we find innumerable papers in which the existence of a stellar evolution is assumed but its nature remains hidden; in which the Kelvin-Helmholtz gravitational contraction was considered the only possible stellar energy source, and nuclear energy remained entirely unanticipated. But at the same time, and sometimes in the same volume of the *Astrophysical Journal,* there is acknowledgment of curious work being done on radioactivity by a man named Becquerel in France. Here again we see the two apparently unrelated threads moving through our few-years snapshot of late-nineteenth-century astronomy and destined to intertwine forty years later.

There are many related examples—for instance, in the interpretation of series spectra of nonhydrogenic elements obtained at the telescope and pursued in the laboratory. New physics and new astronomy were the complementary sides of the emerging science of astrophysics.

Accordingly, it is difficult not to wonder how many of the deepest present debates—for example, on the nature of quasars, or the properties of black holes, or the emission geometry of pulsars—must await an intertwining with new developments in physics. If the experience of seventy-five years ago is any guide, there will already be people today who dimly guess which physics will join with which astronomy. And a few years later, the connection will be considered obvious.

We also see in the nineteenth-century material a number of cases where the observational methods or their interpretations are clearly in default by present standards. Planetary periods deduced to ten significant figures by the comparison of two drawings made by different people of features we now know to be unreal to begin with is one of the worst examples. But there are many others, including a plethora of "double-star measurements" of widely separated objects, which are mainly physically unconnected stars; a fascination with pressure and other effects on the frequencies of spectral lines when no one is paying any attention to curve of growth analysis; and acrimonious debates on the presence or absence of some substance based solely upon naked-eye spectroscopy.

Also curious is the sparseness of the physics in late-Victorian astrophysics. Reasonably sophisticated physics is almost exclusively the province of geometrical and physical optics, the photographic process, and celestial mechanics. To make theories of stellar evolution based on stellar spectra without wondering much about the dependence of excitation and ionization on temperature, or attempting to calculate the subsurface temperature of the Moon without ever solving Fourier's equation of heat conduction seems to me to be less than quaint. In seeing elaborate empirical representations of laboratory spectra, the modern reader becomes impatient for Bohr and Schrödinger and their successors to come along and develop quantum mechanics.

I wonder how many of our present debates and most celebrated theories will appear, from the vantage point of the year 2049, marked by shoddy observations, indifferent intellectual powers or inadequate physical insight. I have the sense that we are today more self-critical than scientists were in 1899; that because of the larger population of astronomers, we check each other's results more often; and that, in part because of the existence of organizations like the American Astronomical Society, the standards of exchange and discussion of results have risen significantly. I hope our colleagues of 2049 will agree.

The major advance between 1899 and 1974 must be considered technological. But in 1899 the world's largest refractor had been built. It is still the world's largest refractor. A reflector of 100-inch aperture was beginning to be considered. We have improved on that aperture only by a factor of two in the intervening years. But what would our colleagues of 1899—living after Hertz but before Marconi—have made of the Arecibo Observatory, or the Very Large Array, or Very Long Baseline Interferometry (VLBI)? Or checking out the debate on the period of rotation of Mercury by radar Doppler spectroscopy? Or testing the nature of the lunar surface by returning some of it to Earth? Or pursuing the problem of the nature and habitability of Mars by orbiting it for a year and returning 7,200 photographs of it, each of higher quality than the best 1899 photographs of the Moon? Or landing on the planet with imaging systems, microbiology experimentation, seismometers and gas chromatograph/mass spectrometers, which did not even vaguely exist in 1899? Or testing cosmological models by orbital ultraviolet spectroscopy of interstellar deuterium—when neither the models to be tested nor the existence of the atom that tests it were known in 1899, much less the technique of observation?

It is clear that in the past seventy-five years American and world astronomy has moved enormously beyond even the most romantic speculations of the late-Victorian astronomers. And in the next seventy-five years? It is possible to make pedestrian predictions. We will have completely examined the electromagnetic spectrum from rather short gamma rays to rather long radio waves. We will have sent unmanned spacecraft to all of the planets and most of the satellites in the solar system. We will have launched spacecraft into the Sun to do experimental stellar structure, beginning perhaps—because of the low temperatures—with the sunspots. Hale would have appreciated that. I think it possible that seventy-five years from now, we will have launched subrelativistic spacecraft—traveling at about 0.1 the speed of light—to the nearby stars. Among other benefits, such missions would permit direct examination of the interstellar medium and give us a longer baseline for VLBI than many are thinking of today. We will have to invent some new superlative to succeed "very" —perhaps "ultra." The nature of pulsars, quasars and black holes should by then be well in hand, as well as the answers to some of the deepest cosmological questions. It is even possible that we will have opened up a regular communications channel with civilizations on planets of other stars, and that the cutting edge of astronomy as well as many other sciences will come from a kind of *Encyclopaedia Galactica,* transmitted at very high bit rates to some immense array of radio telescopes.

But in reading the astronomy of seventy-five years ago, I think it likely that, except for interstellar contact, these achievements, while interesting, will be considered rather old-fashioned astronomy, and that the real frontiers and the fundamental excitement of the science will be in areas that depend on new physics and new technology, which we can today at best dimly glimpse.

CHAPTER 22

THE QUEST FOR EXTRATERRESTRIAL INTELLIGENCE

◆◆◆

> Now the Sirens have a still more fatal weapon than their song, namely their silence . . . Someone might possibly have escaped from their singing; but from their silence, certainly never.
>
> FRANZ KAFKA,
> *Parables*

THROUGH ALL of our history we have pondered the stars and mused whether humanity is unique or if, somewhere else in the dark of the night sky, there are other beings who contemplate and wonder as we do, fellow thinkers in the cosmos. Such beings might view themselves and the universe differently. Somewhere else there might be very exotic biologies and technologies and societies. In a cosmic setting vast and old beyond ordinary human understanding, we are a little lonely; and we ponder the ultimate significance, if any, of our tiny but exquisite blue planet. The search for extraterrestrial intelligence is the search for a generally acceptable cosmic context for the human species. In the deepest sense, the search for extraterrestrial intelligence is a search for ourselves.

In the last few years—in one-millionth the lifetime of our species on this planet—we have achieved an extraordinary technological capability which enables us to seek out unimaginably distant civilizations even if they are no more advanced than we. That capability is called radio astronomy and involves single radio telescopes, collections or arrays of radio telescopes, sensitive radio detectors, advanced computers for processing received data, and the imagination and skill of dedicated scientists. Radio astronomy has in the last decade opened a new window on the physical universe. It may also, if we are wise enough to make the effort, cast a profound light on the biological universe.

Some scientists working on the question of extraterrestrial intelligence, myself among them, have attempted to estimate the number of advanced technical civilizations—defined operationally as societies capable of radio astronomy—in the Milky Way Galaxy. Such estimates are little better than guesses. They require assigning numerical values to quantities such as the numbers and ages of stars; the abundance of planetary systems and the likelihood of the origin of life, which we know less well; and the probability of the evolution of intelligent life and the lifetime of technical civilizations, about which we know very little indeed.

When we do the arithmetic, the sorts of numbers we come up with are, characteristically, around a million technical civilizations. A million civilizations is a breathtakingly large number, and it is exhilarating to imagine the diversity, lifestyles and commerce of those million worlds. But the Milky Way Galaxy contains some 250 billion stars, and even with a million civilizations, less than one star in 200,000 would have a planet inhabited by an advanced civilization. Since we have little idea which stars are likely candidates, we will have to examine a very large number of them. Such considerations suggest that the quest for extraterrestrial intelligence may require a significant effort.

Despite claims about ancient astronauts and unidentified flying objects, there is no firm evidence for past visitations of the Earth by other civilizations (see Chapters 5 and 6). We are restricted to remote signaling and, of the long-distance techniques available to our technology, radio is by far the best. Radio telescopes are relatively inexpensive; radio signals travel at the speed of light, faster than which nothing can go; and the use of radio for communication is not a short-sighted or anthropocentric activity. Radio represents a large part of the electromagnetic spectrum, and any technical civilization anywhere in the Galaxy will have discovered radio early—just as in the last few centuries we have explored the entire electromagnetic spectrum from short gamma rays to very long radio waves. Advanced civilizations might very well use some other means of communication with their peers. But if they wish to communicate with backward or emerging civilizations, there are only a few obvious methods, the chief of which is radio.

The first serious attempt to listen for possible radio signals from other civilizations was carried out at the National Radio Astronomy Observatory in Greenbank, West Virginia, in 1959 and 1960. It was organized by Frank Drake, now at Cornell University, and was called Project Ozma, after the princess of the Land of Oz, a place very exotic, very distant and very difficult to reach. Drake examined two nearby stars, Epsilon Eridani and Tau Ceti, for a few weeks with negative results. Positive results would have been astonishing because as we have seen, even rather optimistic estimates of the number of technical civilizations in the Galaxy imply that several hundred thousand stars must be examined in order to achieve success by random stellar selection.

Since Project Ozma, there have been six or eight other such programs, all at a rather modest level, in the United States, Canada and the Soviet Union. All results have been negative. The total number of individual stars examined to date in this way is less than a thousand. We have performed something like one tenth of one percent of the required effort.

However, there are signs that much more serious

efforts may be mustered in the reasonably near future. All the observing programs to date have involved quite tiny amounts of time on large telescopes, or when large amounts of time have been committed, only very small radio telescopes could be used. A comprehensive examination of the problem was recently made by a NASA committee chaired by Philip Morrison of the Massachusetts Institute of Technology. The committee identified a wide range of options, including new (and expensive) giant ground-based and spaceborne radio telescopes. It also pointed out that major progress can be made at modest cost by the development of more sensitive radio receivers and of ingenious computerized data-processing systems. In the Soviet Union there is a state commission devoted to organizing a search for extraterrestrial intelligence, and the large RATAN-600 radio telescope in the Caucasus, recently completed, is devoted part-time to this effort. Hand in hand with the recent spectacular advances in radio technology, there has been a dramatic increase in the scientific and public respectability of the entire subject of extraterrestrial life. A clear sign of the new attitude is the Viking missions to Mars, which are to a significant extent dedicated to the search for life on another planet.

But along with the burgeoning dedication to a serious search, a slightly negative note has emerged which is nevertheless very interesting. A few scientists have lately asked a curious question: If extraterrestrial intelligence is abundant, why have we not already seen its manifestations? Think of the advances by our own technical civilization in the past ten thousand years and imagine such advances continued over millions or billions of years more. If only a tiny fraction of advanced civilizations are millions or billions of years more advanced than ours, why have they not produced artifacts, devices or even industrial pollution of such magnitude that we would have detected it? Why have they not restructured the entire Galaxy for their convenience?

Skeptics also ask why there is no clear evidence of extraterrestrial visits to Earth. We have already launched slow and modest interstellar spacecraft. A society more advanced than ours should be able to ply the spaces between the stars conveniently if not effortlessly. Over millions of years such societies should have established colonies, which might themselves launch interstellar expeditions. Why are they not here? The temptation is to deduce that there are at most a few advanced extraterrestrial civilizations—either because statistically we are one of the first technical civilizations to have emerged or because it is the fate of all such civilizations to destroy themselves before they are much further along than we.

It seems to me that such despair is quite premature. All such arguments depend on our correctly surmising the intentions of beings far more advanced than ourselves, and when examined more closely I think these arguments reveal a range of interesting human conceits. Why do we expect that it will be easy to recognize the manifestations of very advanced civilizations? Is our situation not closer to that of members of an isolated society in the Amazon basin, say, who lack the tools to detect the powerful international radio and television traffic that is all around them? Also, there is a wide range of incompletely understood phenomena in astronomy. Might the modulation of pulsars or the energy source of quasars, for example, have a technological origin? Or perhaps there is a galactic ethic of noninterference with backward or emerging civilizations. Perhaps there is a waiting time before contact is considered appropriate, so as to give us a fair opportunity to destroy ourselves first, if we are so inclined. Perhaps all societies significantly more advanced than our own have achieved an effective personal immortality and lose the motivation for interstellar gallivanting, which may, for all we know, be a typical urge only of adolescent civilizations. Perhaps mature civilizations do not wish to pollute the cosmos. There is a very long list of such "perhapses," few of which we are in a position to evaluate with any degree of assurance.

The question of extraterrestrial civilizations seems to me entirely open. Personally, I think it far more difficult to understand a universe in which we are the only technological civilization, or one of a very few, than to conceive of a cosmos brimming over with intelligent life. Many aspects of the problem are, fortunately, amenable to experimental verification. We can search for planets of other stars, seek simple forms of life on such nearby planets as Mars, and perform more extensive laboratory studies on the chemistry of the origin of life. We can investigate more deeply the evolution of organisms and societies. The problem cries out for a long-term, open-minded, systematic search, with nature as the only arbiter of what is or is not likely.

If there are a million technical civilizations in the Milky Way Galaxy, the average separation between civilizations is about 300 light-years. Since a light-year is the distance that light travels in one year (a little under 6 trillion miles), this implies that the one-way transit time for an interstellar communication from the nearest civilization is some 300 years. The time for a query and a response would be 600 years. This is the reason that interstellar dialogues are much less likely—particularly around the time of first contact—than interstellar monologues. At first sight, it seems remarkably selfless that a civilization might broadcast radio messages with no hope of knowing, at least in the immediate future, whether they have been received and what the response to them might be. But human beings often perform very similar actions as, for example, burying time capsules to be recovered by future generations, or even writing books, composing music and creating art intended for posterity. A civilization that had been aided by the receipt of such a message in its past might wish similarly to benefit other emerging technical societies.

For a radio search program to succeed, the Earth must be among the intended beneficiaries. If the transmitting civilization were only slightly more advanced than we are, it would possess ample radio power for interstellar communication—so much, perhaps, that the broadcasting could be delegated to relatively small groups of radio hobbyists and partisans of primitive civilizations. If an entire planetary government or an alliance of worlds carried out the project, the broadcasters could transmit to a very large number of stars, so large that a message is likely to be beamed our way, even though there may be no reason to pay special attention to our region of the sky.

It is easy to see that communication is possible, even without any previous agreement or contact between transmitting and receiving civilizations. There is no difficulty in envisioning an interstellar radio message that unambiguously arises from intelligent life. A modulated signal (beep, beep-beep, beep-beep-beep . . .) comprising the numbers 1, 2, 3, 5, 7, 11, 13, 17, 19, 23, 29, 31—the first dozen prime numbers—could have only a biological origin. No prior agreement between civilizations and no precautions against Earth chauvinism are required to make this clear.

Such a message would be an announcement, or beacon signal, indicating the presence of an advanced civilization but communicating very little about its nature. The beacon signal might also note a particular frequency where the main message is to be found, or might indicate that the principal message can be found at higher time resolution at the frequency of the beacon signal. The communication of quite complex information is not very difficult, even for civilizations with extremely different biologies and social conventions. Arithmetical statements can be transmitted, some true and some false, each followed by an appropriate coded word (in dahs and dits, for example), which would transmit the ideas of true and false, concepts that many people might guess would be extremely difficult to communicate in such a context.

But by far the most promising method is to send pictures. A repeated message that is the product of two prime numbers is clearly to be decoded as a two-dimensional array, or raster—that is, a picture. The product of three prime numbers might be a three-dimensional still picture or one frame of a two-dimensional motion picture. As an example of such a message, consider an array of zeros and ones which could be long and short beeps or tones on two adjacent frequencies, or tones of different amplitudes, or even signals with different radio polarizations. In 1974 such a message was transmitted to space from the 305-meter antenna at the Arecibo Observatory in Puerto Rico, which Cornell University runs for the National Science Foundation. The occasion was a ceremony marking the resurfacing of the Arecibo dish, the largest radio/radar telescope on the planet Earth. The signal was sent to a collection of stars called M13, a globular cluster comprising about a million separate suns which happened to be overhead at the time of the ceremony. Since M13 is 24,000 light-years away, the message will take 24,000 years to arrive there. If any responsive creature is listening, it will be 48,000 years before we receive a reply. The Arecibo message was clearly intended not as a serious attempt at interstellar communication, but rather as an indication of the remarkable advances in terrestrial radio technology.

The decoded message says something like this: "Here is how we count from one to ten. Here are the atomic numbers of five chemical elements—hydrogen, carbon, nitrogen, oxygen and phosphorus—that we think are interesting or important. Here are some ways to put these atoms together: the molecules adenine, thymine, guanine and cytosine, and a chain composed of alternating sugars and phosphates. These molecular building blocks are in turn put together to form a long molecule of DNA comprising about four billion links in the chain. The molecule is a double helix. In some way this molecule is important for the clumsy-looking creature at the center of the message. That creature is 14 radio wavelengths, or about 176 centimeters, high. There are about four billion of these creatures on the third planet from our star. There are nine planets altogether—four little ones on the inside, four big ones toward the outside and one little one at the extremity. This message is brought to you courtesy of a radio telescope 2,430 wavelengths, or 306 meters, in diameter. Yours truly."

With many similar pictorial messages, each consistent with and corroborating the others, it is very likely that almost unambiguous interstellar radio communication could be achieved even between two civilizations that have never met. Our immediate objective is not to send such messages because we are very young and backward; we wish to listen.

The detection of intelligent radio signals from the depths of space would approach in an experimental and scientifically rigorous manner many of the most profound questions that have concerned scientists and philosophers since prehistoric times. Such a signal would indicate that the origin of life is not an extraordinary, difficult or unlikely event. It would imply that, given billions of years for natural selection, simple forms of life evolve generally into complex and intelligent forms, as on Earth; and that such intelligent forms commonly produce an advanced technology, as has also occurred here. But it is not likely that the transmissions we receive will be from a society at our own level of technological advance. A society only a little more backward than ours will not have radio astronomy at all. The most likely case is that the message will be from a civilization far in our technological future. Thus, even before we decode such a message, we will have gained an invaluable piece of knowledge: that it is possible to avoid the dangers of the period through which we are now passing.

There are some who look on our global problems here on Earth—at our vast national antagonisms, our nuclear arsenals, our growing populations, the disparity between the poor and the affluent, shortages of food and resources, and our inadvertent alterations of the natural environment—and conclude that we live in a system that has suddenly become unstable, a system that is destined soon to collapse. There are others who believe that our problems are soluble, that humanity is still in its childhood, that one day soon we will grow up. The receipt of a single message from space would show that it is possible to live through such technological adolescence: the transmitting civilization, after all, has survived. Such knowledge, it seems to me, might be worth a great price.

Another likely consequence of an interstellar message is a strengthening of the bonds that join all human and other beings on our planet. The sure lesson of evolution is that organisms elsewhere must have separate evolutionary pathways; that their chemistry and biology and very likely their social organizations will be profoundly dissimilar to anything on Earth. We may well be able to communicate with them because we share a common universe—because the laws of physics and chemistry and the regularities of astronomy are universal. But they may always be, in the deepest sense, different. And in the face of this difference, the animosities that divide the peoples of the Earth may wither. The differences among human beings of separate races and nationalities, religions and sexes, are likely to be insignificant compared to the differences between all human and all extraterrestrial intelligent beings.

If the message comes by radio, both transmitting and receiving civilizations will have in common at least a knowledge of radiophysics. The commonality of the physical sciences is the reason that many scientists expect the messages from extraterrestrial civilizations to be decodable—probably in a slow and halting manner, but unambiguously nevertheless. No one is wise enough to predict in detail what the consequences of such a decoding will be, because no one is wise enough to understand beforehand what the nature of the message will be. Since the transmission is likely to be from a civilization far in advance of our own, stunning insights are possible in the physical, biological and social sciences, in the novel perspective of a quite different kind of intelligence. But decoding will probably be a task of years and decades.

Some have worried that a message from an advanced society might make us lose faith in our own, might deprive us of the initiative to make new discoveries if it seemed that others had made those discoveries already, or might have other negative consequences. This is rather like a student dropping out of school because his teachers and textbooks are more learned than he is. We are free to ignore an interstellar message if we find it offensive. If we choose not to respond, there is no way for the transmitting civilization to determine that its message was received and understood on the tiny distant planet Earth. The translation of a radio message from the depths of space, about which we can be as slow and cautious as we wish, seems to pose few dangers to mankind; instead, it holds the greatest promise of both practical and philosophical benefits.

In particular, it is possible that among the first contents of such a message may be detailed prescriptions for the avoidance of technological disaster, for a passage through adolescence to maturity. Perhaps the transmissions from advanced civilizations will describe which pathways of cultural evolution are likely to lead to the stability and longevity of an intelligent species, and which other paths lead to stagnation or degeneration or disaster. There is, of course, no guarantee that such would be the contents of an interstellar message, but it would be foolhardy to overlook the possibility. Perhaps there are straightforward solutions, still undiscovered on Earth, to problems of food shortages, population growth, energy supplies, dwindling resources, pollution and war.

While there will surely be differences among civilizations, there may well be laws of development of civilizations which cannot be glimpsed until information is available about the evolution of many civilizations. Because of our isolation from the rest of the cosmos, we have information on the evolution of only one civilization—our own. And the most important aspect of that evolution—the future—remains closed to us. Perhaps it is not likely, but it is certainly possible that the future of human civilization depends on the receipt and decoding of interstellar messages from extraterrestrial civilizations.

And what if we make a long-term, dedicated search for extraterrestrial intelligence and fail? Even then we surely will not have wasted our time. We will have developed an important technology, with applications to many other aspects of our own civilization. We will have added greatly to our knowledge of the physical universe. And we will have calibrated something of the importance and uniqueness of our species, our civilization and our planet. For if intelligent life is scarce or absent elsewhere, we will have learned something significant about the rarity and value of our culture and our biological patrimony, painstakingly extracted over 4.6 billion years of tortuous evolutionary history. Such a finding will stress, as perhaps nothing else can, our responsibilities to the dangers of our time: because the most likely explanation of negative results, after a comprehensive and resourceful search, is that societies commonly destroy themselves before they are advanced enough to establish a high-power radio-transmitting service. In an interesting sense, the organization of a search for interstellar radio messages, quite apart from the outcome, is likely to have a cohesive and constructive influence on the whole of the human predicament.

But we will not know the outcome of such a search, much less the contents of messages from interstellar civilizations, if we do not make a serious effort to listen for signals. It may be that civilizations are divided into two great classes: those that make such an effort, achieve contact and become new members of a loosely tied federation of galactic communities, and those that cannot or choose not to make such an effort, or who lack the imagination to try, and who in consequence soon decay and vanish.

It is difficult to think of another enterprise within our capability and at a relatively modest cost that holds as much promise for the future of humanity.

PART V

ULTIMATE QUESTIONS

CHAPTER 23

A SUNDAY SERMON

◆◆◆

Extinguished theologians lie about the cradle of every science as the strangled snakes beside [the cradle] of Hercules.

T. H. HUXLEY (1860)

We have seen the highest circle of spiraling powers. We have named this circle God. We might have given it any other name we wished: Abyss, Mystery, Absolute Darkness, Absolute Light, Matter, Spirit, Ultimate Hope, Ultimate Despair, Silence.

NIKOS KAZANTZAKIS (1948)

THESE DAYS, I often find myself giving scientific talks to popular audiences. Sometimes I am asked to discuss planetary exploration and the nature of the other planets; sometimes, the origin of life or intelligence on Earth; sometimes, the search for life elsewhere; and sometimes, the grand cosmological perspective. Since I have, more or less, heard these talks before, the question period holds my greatest interest. It reveals the attitudes and concerns of people. The most common questions asked are on unidentified flying objects and ancient astronauts—what I believe are thinly disguised religious queries. Almost as common—particularly after a lecture in which I discuss the evolution of life or intelligence—is: "Do you believe in God?" Because the word "God" means many things to many people, I frequently reply by asking what the questioner means by "God." To my surprise, this response is often considered puzzling or unexpected: "Oh, you know, *God*. Everyone knows who God is." Or "Well, kind of a force that is stronger than we are and that exists everywhere in the universe." There are a number of such forces. One of them is called gravity, but it is not often identified with God. And not everyone does know what is meant by "God." The concept covers a wide range of ideas. Some people think of God as an outsized, light-skinned male with a long white beard, sitting on a throne somewhere up there in the sky, busily tallying the fall of every sparrow. Others—for example, Baruch Spinoza and Albert Einstein—considered God to be essentially the sum total of the physical laws which describe the universe. I do not know of any compelling evidence for anthropomorphic patriarchs controlling human destiny from some hidden celestial vantage point, but it would be madness to deny the existence of physical laws. Whether we believe in God depends very much on what we mean by God.

In the history of the world there have been, probably, tens of thousands of different religions. There is a well-intentioned pious belief that they are all fundamentally identical. In terms of an underlying psychological resonance, there may indeed be important similarities at the cores of many religions, but in the details of ritual and doctrine, and the *apologias* considered to be authenticating, the diversity of organized religions is striking. Human religions are mutually exclusive on such fundamental issues as one god versus many; the origin of evil; reincarnation; idolatry; magic and witchcraft; the role of women; dietary proscriptions; rites of passage; ritual sacrifice; direct or mediated access to deities;

slavery; intolerance of other religions; and the community of beings to whom special ethical considerations are due. We do no service to religion in general or to any doctrine in particular if we paper over these differences. Instead, I believe we should understand the world views from which differing religions derive and seek to understand what human needs are fulfilled by those differences.

Bertrand Russell once told of being arrested because he peacefully protested Britain's entry into World War I. The jailer asked—then a routine question for new arrivals—Russell's religion. Russell replied, "Agnostic," which he was asked to spell. The jailer smiled benignly, shook his head and said, "There's many different religions, but I suppose we all worship the same God." Russell commented that the remark cheered him for weeks. And there may not have been much else to cheer him in that prison, although he did manage to write the entire *Introduction to Mathematical Philosophy* and started reading for his work *The Analysis of Mind* within its confines.

Many of the people who ask whether I believe in God are requesting reassurance that their particular belief system, whatever it is, is consistent with modern scientific knowledge. Religion has been scarred in its confrontation with science, and many people—but by no means all—are reluctant to accept a body of theological belief that is too obviously in conflict with what else we know. Apollo 8 accomplished the first manned lunar circumnavigation. In a more or less spontaneous gesture, the Apollo 8 astronauts read from the first verse of the Book of Genesis, in part, I believe, to reassure the taxpayers back in the United States that there were no real inconsistencies between conventional religious outlooks and a manned flight to the Moon. Orthodox Muslims, on the other hand, were outraged after Apollo 11 astronauts accomplished the first manned lunar landing, because the Moon has a special and sacred significance in Islam. In a different religious context, after Yuri Gagarin's first orbital flight, Nikita Khrushchev, the chairman of the Council of Ministers of the USSR, noted that Gagarin had stumbled on no gods or angels up there—that is, Khrushchev reassured his audience that manned orbital flight was not inconsistent with its beliefs.

In the 1950s a Soviet technical journal called *Voprosy Filosofii* (Problems in Philosophy) published an article that argued—very unconvincingly, it seemed to me—that dialectical materialism required there to be life on every planet. Some time later an agonized official rebuttal appeared, decoupling dialectical materialism from exobiology. A clear prediction in an area undergoing vigorous study permits doctrines to be subject to disproof. The last posture a bureaucratic religion wishes to find itself in is vulnerability to disproof, where an experiment can be performed on which the religion stands or falls. And so the fact that life has not been found on the Moon has left the foundations of dialectical materialism unshaken. Doctrines that make no predictions are less compelling than those which make correct predictions; they are in turn more successful than doctrines that make false predictions.

But not always. One prominent American religion confidently predicted that the world would end in 1914. Well, 1914 has come and gone, and—while the events of that year were certainly of some importance—the world does not, at least so far as I can see, seem to have ended. There are at least three responses that an organized religion can make in the face of such a failed and fundamental prophecy. They could have said, "Oh, did we say '1914'? So sorry, we meant '2014.' A slight error in calculation. Hope you weren't inconvenienced in any way." But they did not. They could have said, "Well, the world *would* have ended, except we prayed very hard and interceded with God so He spared the Earth." But they did not. Instead, they did something much more ingenious. They announced that the world *had* in fact ended in 1914, and if the rest of us hadn't noticed, that was our lookout. It is astonishing in the face of such transparent evasions that this religion has any adherents at all. But religions are tough. Either they make no contentions which are subject to disproof or they quickly redesign doctrine after disproof. The fact that religions can be so shamelessly dishonest, so contemptuous of the intelligence of their adherents, and still flourish does not speak very well for the tough-mindedness of the believers. But it does indicate, if a demonstration were needed, that near the core of the religious experience is something remarkably resistant to rational inquiry.

Andrew Dickson White was the intellectual guiding light, founder and first president of Cornell University. He was also the author of an extraordinary book called *The Warfare of Science with Theology in Christendom,* considered so scandalous at the time it was published that his co-author requested his name omitted. White was a man of substantial religious feeling.* But he outlined the long and painful history of erroneous claims which religions had made about the nature of the world, and how, when people directly investigated the nature of the world and discovered it to be different from doctrinal contentions, such people were persecuted and their ideas suppressed. The aged Galileo was threatened by the Catholic hierarchy with torture because he proclaimed the Earth to move. Spinoza was excommunicated by the Jewish hierarchy, and there is hardly an organized religion with a firm body of doctrine which has not at one time or another persecuted people for the crime of open inquiry. Cornell's own devotion to free and non-sectarian inquiry was considered so objectionable in the last quarter of the nineteenth century that ministers advised high school graduates that it was better to receive no college education than to attend so impious an institution. Indeed, this Sage Chapel was constructed in part to placate the pious—although, I am glad to say, it has from time to time made serious efforts at open-minded ecumenicism.

---

* White seems also to have been responsible for the exemplary custom of not awarding honorary doctoral degrees at Cornell University: he was concerned about a potential abuse, that honorary degrees would be traded for financial gifts and bequests. White was a man of strong and courageous ethical standards.

Many of the controversies which White describes are about origins. It used to be believed that every event in the world—the opening of a morning glory, let us say—was due to direct microintervention by the Deity. The flower was unable to open by itself. God had to say, "Hey, flower, open." The application of this idea to human affairs has often had desultory social consequences. For one thing it seems to imply that we are not responsible for our actions. If the play of the world is produced and directed by an omnipotent and omniscient God, does it not follow that every evil that is perpetrated is God's doing? I know this idea is an embarrassment in the West, and attempts to avoid it include the contention that what seems to be evil is really part of the Divine Plan, too complex for us to fathom; or that God chose to cloud his own vision about the causality skein when he set out to make the world. There is nothing utterly impossible about these philosophical rescue attempts, but they do seem to have very much the character of propping up a teetering ontological structure.* In addition, the idea of microintervention in the affairs of the world has been used to support the established social, political and economic conventions. There was, for example, the idea of a "Divine Right of Kings," seriously argued by philosophers such as Thomas Hobbes. If you had revolutionary thoughts directed, let us say, toward George III, you were guilty of blasphemy and impiety, religious crimes, as well as such more commonplace political crimes as treason.

There are many legitimate scientific issues relating to origins and ends: What is the origin of the human species? Where did plants and animals come from? How

* Many statements about God are confidently made by theologians on grounds that today at least sound specious. Thomas Aquinas claimed to prove that God cannot make another God, or commit suicide, or make a man without a soul, or even make a triangle whose interior angles do not equal 180 degrees. But Bolyai and Lobachevsky were able to accomplish this last feat (on a curved surface) in the nineteenth century, and they were not even approximately gods. It is a curious concept this, of an omnipotent God with a long list of things he is forbidden to do by the fiat of the theologians.

did life arise? the Earth, the planets, the Sun, the stars? Does the universe have an origin, and if so, what? And finally, a still more fundamental and exotic question, which many scientists would say is essentially untestable and therefore meaningless: Why are the laws of nature the way they are? The idea that a God or gods is necessary to effect one or more of these origins has been under repeated attack over the last few thousand years. Because we know something about phototropism and plant hormones, we can understand the opening of the morning glory independent of divine microintervention. It is the same for the entire skein of causality back to the origin of the universe. As we learn more and more about the universe, there seems less and less for God to do. Aristotle's view was of God as an unmoved prime mover, a *roi fainéant,* a do-nothing king who establishes the universe in the first place and then sits back and watches the intricate, intertwined chains of causality course down through the ages. But this seems abstract and removed from everyday experience. It is a little unsettling and pricks at human conceits.

Humans seem to have a natural abhorrence of an infinite regression of causes, and this distaste is at the root of the most famous and most effective demonstrations of the existence of God by Aristotle and Thomas Aquinas. But these thinkers lived before the infinite series was a mathematical commonplace. If the differential and integral calculus or transfinite arithmetic had been invented in Greece in the fifth century B.C., and not subsequently suppressed, the history of religion in the West might have been very different—or at any rate we would have seen less of the pretension that theological doctrine can be convincingly demonstrated by rational argument to those who reject alleged divine revelation, as Aquinas attempted in the *Summa Contra Gentiles.*

When Newton explained the motion of the planets by the universal theory of gravitation, it no longer was necessary for angels to push and pummel the planets about. When Pierre Simon, the Marquis de Laplace, proposed to explain the origin of the solar system—although not the origin of matter—in terms of physical laws as well, even the necessity for a god involved in the origins of things seemed profoundly challenged. Laplace is said to have presented an edition of his seminal mathematical work *Mécanique céleste* to Napoleon aboard ship in the Mediterranean during the Napoleonic expedition to Egypt, 1798 to 1799. A few days later, so the story goes, Napoleon complained to Laplace that he had found no mention of God in the text.* Laplace's response has been recorded: "Sire, I have no need of that hypothesis." The idea of God as a hypothesis rather than as an obvious truth is by and large a modern idea in the West—although it was certainly discussed seriously and wryly by the Ionian philosophers of 2,400 years ago.

It is often considered that at least the origin of the universe requires a God—indeed, an Aristotelian idea.† This is a point worth looking at in a little more detail. First of all, it is perfectly possible that the universe is infinitely old and therefore requires no Creator. This is consistent with existing knowledge of cosmology, which permits an oscillating universe in which the events since the Big Bang are merely the latest incarnation in an infinite series of creations and destructions of the universe. But secondly, let us consider the idea of a universe created somehow from nothing by God. The question

---

* It is a charming notion that Napoleon actually spent his days aboard ship perusing the highly mathematical *Mécanique céleste.* But he was seriously interested in science and made an earnest attempt to survey the latest findings (see *The Society of Arcueil: A View of French Science at the Time of Napoleon I* by Maurice Crosland, Cambridge, Harvard University Press, 1967). Napoleon did not pretend to read all of the *Mécanique céleste* and wryly wrote to Laplace on another occasion, "The first six months which I can spare will be employed in reading it." But he also remarked, on another of Laplace's books, "Your works contribute to the glory of the nation. The progress and perfection of mathematics are linked closely with the prosperity of the state."

† However, from astronomical arguments Aristotle concluded that there were several dozen unmoved prime movers in the universe. Aristotelian arguments for a prime mover would seem to have polytheistic consequences that might be considered dangerous by contemporary Western theologians.

naturally arises—and many ten-year-olds spontaneously think of it before being discouraged by their elders—where does God come from? If we answer that God is infinitely old or present simultaneously in all epochs, we have solved nothing, except perhaps verbally. We have merely postponed by one step coming to grips with the problem. A universe that is infinitely old and a God that is infinitely old are, I think, equally deep mysteries. It is not readily apparent why one should be considered more reliably established than the other. Spinoza might have said that the two possibilities are not really different ideas at all.

I think it is wise, when coming face to face with such profound mysteries, to feel a little humility. The idea that scientists or theologians, with our present still puny understanding of this vast and awesome cosmos, can comprehend the origins of the universe is only a little less silly than the idea that Mesopotamian astronomers of 3,000 years ago—from whom the ancient Hebrews borrowed, during the Babylonian captivity, the cosmological accounts in the first chapter of Genesis—could have understood the origins of the universe. We simply do not know. The Hindu holy book, the Rig Veda (X:129), has a much more realistic view of the matter:

> Who knows for certain? Who shall here declare it?
> Whence was it born, whence came creation?
> The gods are later than this world's formation;
> Who then can know the origins of the world?
> None knows whence creation arose;
> And whether he has or has not made it;
> He who surveys it from the lofty skies,
> Only he knows—or perhaps he knows not.

But the times we live in are very interesting ones. Questions of origins, including some questions relating to the origin of the universe, may in the next few decades be amenable to experimental inquiry. There is no conceivable answer to the grand cosmological questions which will not resonate with the religious sensibilities of human beings. But there is a chance that the answers will discomfit a great many bureaucratic and doctrinal religions. The idea of religion as a body of belief, immune to criticism, fixed forever by some founder is, I think, a prescription for the long-term decay of the religion, especially lately. In questions of origins and ends, the religious and the scientific sensibilities have much the same objectives. Human beings are built in such a way that we passionately wish to answer these questions—perhaps because of the mystery of our own individual origins. But our contemporary scientific insights, while limited, are much deeper than those of our Babylonian predecessors of 1,000 B.C. Religions unwilling to accommodate to change, both scientific and social, are, I believe, doomed. A body of belief cannot be alive and relevant, vibrant and growing, unless it is responsive to the most serious criticism that can be mustered against it.

The First Amendment to the United States Constitution encourages a diversity of religions but does not prohibit criticism of religion. In fact it protects and encourages criticism of religion. Religions ought to be subject to at least the same degree of skepticism as, for example, contentions about UFO visitations or Velikovskian catastrophism. I think it is healthy for the religions themselves to foster skepticism about the fundamental underpinnings of their evidential bases. There is no question that religion provides a solace and support, a bulwark in time of emotional need, and can serve extremely useful social roles. But it by no means follows that religion should be immune from testing, from critical scrutiny, from skepticism. It is striking how little skeptical discussion of religion there is in the nation that Tom Paine, the author of *The Age of Reason,* helped to found. I hold that belief systems that cannot survive scrutiny are probably not worth having. Those that do survive scrutiny probably have at least important kernels of truth within them.

Religion used to provide a generally accepted understanding of our place in the universe. That surely has been one of the major objectives of myth and legend, philosophy and religion, as long as there have been human beings. But the mutual confrontation of differing religions and of religion with science has eroded those traditional views, at least in the minds of many.* The way to find out about our place in the universe is by examining the universe and by examining ourselves —without preconceptions, with as unbiased a mind as we can muster. We cannot begin with an entirely clean slate, since we arrive at this problem with predispositions of hereditary and environmental origin; but, after understanding such built-in biases, is it not possible to pry insights from nature?

Proponents of doctrinal religions—ones in which a particular body of belief is prized and infidels scorned—will be threatened by the courageous pursuit of knowledge. We hear from such people that it may be dangerous to probe too deeply. Many people have inherited their religion like their eye color: they consider it not a thing to think very deeply about, and in any case beyond our control. But those with a set of beliefs they profess to feel deeply about, which they have selected without an unbiased sifting through the facts and the alternatives, will feel uncomfortably challenged by searching questions. Anger at queries about our beliefs

* This subject is rich in irony. Augustine was born in Africa in 354 A.D. and in his early years was a Manichean, an adherent of a dualistic view of the universe in which good and evil are in conflict on roughly equal terms, and which was later condemned as a "heresy" by Christian orthodoxy. The possibility that all was not right with Manicheanism occurred to Augustine when he was studying its astronomy. He discovered that even the leading figures in the faith could not justify its murky astronomical notions. This contradiction between theology and science on matters astronomical was the initial impetus moving him toward Catholicism, the religion of his mother, which in later centuries persecuted scientists such as Galileo for trying to improve our understanding of astronomy. Augustine later became Saint Augustine, one of the major intellectual figures in the history of the Roman Catholic church, and his mother became Saint Monica, after whom a suburb of Los Angeles is named. Bertrand Russell wondered what Augustine's view of the conflict between astronomy and theology would have been had he lived in the time of Galileo.

is the body's warning signal: here lies unexamined and probably dangerous doctrinal baggage.

Christianus Huygens wrote a remarkable book around 1670 in which bold and prescient speculations were made about the nature of the other planets in the solar system. Huygens was well aware that there were those who held such speculations and his astronomical observations objectionable: "But perhaps they'll say," Huygens mused, "it does not become us to be so curious and inquisitive in these Things which the Supreme Creator seems to have kept for his own Knowledge: For since he has not been pleased to make any farther Discovery or Revelation of them, it seems little better than presumption to make any inquiry into that which he has thought fit to hide. But these Gentlemen must be told," Huygens then thundered, "that they take too much upon themselves when they pretend to appoint how far and no farther Men shall go in their Searches, and to set bounds to other Mens Industry; as if they knew the Marks that God has placed to Knowledge: or as if Men were able to pass those Marks. If our Forefathers had been at this rate scrupulous, we might have been ignorant still of the Magnitude and Figure of the Earth, or that there was such a place as America."

If we look at the universe in the large, we find something astonishing. First of all, we find a universe that is exceptionally beautiful, intricately and subtly constructed. Whether our appreciation of the universe is because we are a part of that universe—whether, no matter how the universe were put together, we would have found it beautiful—is a proposition to which I do not pretend to have an answer. But there is no question that the elegance of the universe is one of its most remarkable properties. At the same time, there is no question that there are cataclysms and catastrophes occurring regularly in the universe and on the most awesome scale. There are, for example, quasar explosions which probably decimate the nuclei of galaxies. It seems likely that every time a quasar explodes, more than a million worlds are obliterated and countless forms of life, some of them intelligent, are utterly destroyed. This is not the traditional benign universe of conventional religiosity in the West, constructed for the benefit of living and especially of human beings. Indeed, the very scale of the universe—more than a hundred billion galaxies, each containing more than a hundred billion stars—speaks to us of the inconsequentiality of human events in the cosmic context. We see a universe simultaneously very beautiful and very violent. We see a universe that does not exclude a traditional Western or Eastern god, but that does not require one either.

My deeply held belief is that if a god of anything like the traditional sort exists, our curiosity and intelligence are provided by such a god. We would be unappreciative of those gifts (as well as unable to take such a course of action) if we suppressed our passion to explore the universe and ourselves. On the other hand, if such a traditional god does not exist, our curiosity and our intelligence are the essential tools for managing our survival. In either case, the enterprise of knowledge is consistent with both science and religion, and is essential for the welfare of the human species.

CHAPTER 24

GOTT AND THE TURTLES

◆◆◆

Now entertain conjecture of a time
When creeping murmur and the poring dark
Fills the wide vessel of the universe.

WILLIAM SHAKESPEARE,
Henry V, Act IV, Prologue

IN THE EARLIEST myths and legends of our species, there is a common and understandable view of the cosmos: it is anthropocentric. There were gods, to be sure. But the gods had feelings and weaknesses and were very human. Their behavior was seen as capricious. They could be propitiated by sacrifice and prayer. They intervened regularly in human affairs. Various factions of gods supported opposing sides in human warfare. The *Odyssey* expresses a generally held view that it is wise to be kind to strangers: they may be gods in disguise. Gods mate with humans, and the offspring are generally indistinguishable, at least in appearance, from people. The gods live on mountains or in the sky, or in some subterranean or submarine realm—at any rate, far off. It was difficult unambiguously to come upon a god, and so it was hard to check a story told about the gods. Sometimes *their* actions were controlled by more powerful beings yet, as the Fates controlled the Olympian gods. The nature of the universe as a whole, its origin and fate, were not considered well understood. In the Vedic myths, doubt is raised not only about whether the gods created the world but even about whether the gods know who *did* create it. Hesiod in his "Cosmogony" says that the universe was created from (or maybe by) Chaos—perhaps only a metaphor for the difficulty of the problem.

Some ancient Asian cosmological views are close to the idea of an infinite regression of causes, as exemplified in the following apocryphal story: A Western traveler encountering an Oriental philosopher asks him to describe the nature of the world:

"It is a great ball resting on the flat back of the world turtle."

"Ah yes, but what does the world turtle stand on?"

"On the back of a still larger turtle."

"Yes, but what does he stand on?"

"A very perceptive question. But it's no use, mister; it's turtles all the way down."

We now know that we live on a tiny dust mote in an immense and humbling universe. The gods, if they exist, no longer intervene daily in human affairs. We do not live in an anthropocentric universe. And the nature, origin and fate of the cosmos seem to be mysteries far more profound than they were perceived to be by our remote ancestors.

But the situation is once again changing. Cosmology, the study of the universe as a whole, is becoming an experimental science. Information obtained by optical and radio telescopes on the ground, by ultraviolet and X-ray telescopes in Earth orbit, by the measurement of nuclear reactions in laboratories, and by determinations of the abundance of chemical elements in meteorites, is shrinking the arena of permissible cosmological hypotheses; and it is not too much to expect that we will soon have firm observational answers to questions once considered the exclusive preserve of philosophical and theological speculation.

This observational revolution began from an unlikely source. In the second decade of this century there was—as there still is—in Flagstaff, Arizona, an astronomical facility called the Lowell Observatory, established by none other than Percival Lowell, for whom the search for life on other planets was a consuming passion. It was he who popularized and promoted the idea that Mars was crisscrossed with canals, which he believed to be the artifacts of a race of beings enamoured of hydraulic engineering. We now know that the canals do not exist at all. They apparently were the product of wishful thinking and the limitations of observing through the Earth's murky atmosphere.

Among his other interests, Lowell was concerned with the spiral nebulae—exquisite pinwheel-shaped luminous objects in the sky, which we now know to be distant collections of hundreds of billions of individual stars, like the Milky Way Galaxy of which our Sun is a part. But at that time there was no way to determine the distance to these nebulae, and Lowell was interested in an alternative hypothesis—that the spiral nebulae were not enormous, distant, multistellar entities, but rather smaller, closer objects which were the early stages of the condensation of an individual star out of the interstellar gas and dust. As such gas clouds contract under their self-gravitation, the conservation of angular momentum requires that they speed up to rapid rotation and shrink to a thin disc. Rapid rotation can be detected astronomically by spectroscopy, letting light from a distant object pass consecutively through a telescope, a narrow slit and a glass prism or other device which spreads white light out into a rainbow of colors. The spectrum of starlight contains bright and dark lines here and there in the rainbow, images of the slit of the spectrometer. An example is the bright yellow lines emitted by sodium, apparent as we throw a small piece of sodium into a flame. Material made of many different chemical elements will show many different spectral lines. The displacement of these spectral lines from their usual wavelengths when the light source is at rest gives us information on the velocity of the source toward and away from us—a phenomenon called the Doppler effect and familiar to us, in the physics of sound, as the increase or decrease in the pitch of an automobile horn as the car rapidly approaches or recedes.

Lowell is thought to have asked a young assistant, V. M. Slipher, to check the larger spiral nebulae to determine whether one side showed spectral lines shifted toward the red and the other toward the blue, from which it would be possible to deduce the speed of rotation of the nebula. Slipher investigated the spectra of the nearby spiral nebulae but found to his amazement that almost all of them showed a red shift, with virtually no sign of blue shifts anywhere in them. He had found not rotation, but recession. It was as if all the spiral nebulae were retreating from *us*.

A much more extensive set of observations was obtained in the 1920s at the Mount Wilson Observatory by Edwin Hubbell and Milton Humason. Hubbell and Humason developed a method of determining the distance to the spiral nebulae; it became apparent that they were not condensing gas clouds relatively nearby in the Milky Way Galaxy, but themselves great galaxies millions or more light-years away. To their amazement, they also found that the more distant the galaxy, the faster it was receding from us. Since it is unlikely that there is anything special about our position in the cosmos, this is best understood in terms of a general expansion of the universe; all galaxies recede from all others so that an astronomer on any galaxy would observe all other galaxies apparently retreating.

If we extrapolate such a mutual recession back into the past, we find that there was a time—perhaps 15 billion or 20 billion years ago—when all of the galaxies must have been "touching"; that is, confined to an extremely small volume of space. Matter in its present form could not survive such astonishing compressions. The very earliest stages of that expanding universe must have been dominated by radiation rather than matter. It is now conventional to talk of this time as the Big Bang.

Three kinds of explanation have been offered for this expansion of the universe: the Steady State, Big Bang and Oscillating Universe cosmologies. In the Steady State hypothesis, the galaxies recede from one another, the more distant galaxies moving with very high apparent velocities, their light shifted by the Doppler effect to longer and longer wavelengths. There will be a distance at which a galaxy will be moving so fast that it passes over what is called its event horizon and, from our vantage point, disappears. There is a distance so great that, in an expanding universe, there is no chance of getting information from beyond it. As time goes on, if nothing else intervenes, more and more galaxies will disappear over the edge. But in the Steady State cosmology, the matter lost over the edge is exactly compensated for by new matter continuously created everywhere, matter that eventually condenses into new galaxies. With the rate of disappearance of galaxies over the event horizon just balanced by the creation of new galaxies, the universe looks more or less identical from every place and in every epoch. In the Steady State cosmology there is no Big Bang; one hundred billion years ago the universe would have looked just the same, and one hundred billion years from now, likewise. But where does the new matter come from? How can matter be created from nothing? Proponents of the Steady State cosmology answer that it comes from whatever place proponents of the Big Bang get their Bang from. If we can imagine all the matter in the universe discontinuously created from nothing 15 billion to 20 billion years ago, why are we unable to imagine it being created in a tenuous trickle everywhere, continuously and forever? If the Steady State hypothesis is true, there was never a time when the galaxies were much closer. The universe in its largest structures is then unchanging and infinitely old.

But as placid and, in a strange way, as satisfying as the Steady State cosmology is, there is strong evidence against it. Whenever a sensitive radio telescope is pointed anywhere in the sky, the constant chatter of a kind of cosmic static can be detected. The characteristics of this radio noise match almost exactly what we would expect if the early universe was hot and filled with radiation in addition to matter. The cosmic blackbody radiation is very nearly the same everywhere in the sky and looks very much to be the distant rumblings of the Big Bang, cooled and enfeebled by the expansion of the universe but coursing still down the corridors of time. The primeval fireball, the explosive event that initiated the expanding universe, can be observed. Supporters of the Steady State cosmology must now be reduced to positing a large number of special sources of radiation which together somehow mimic exactly the cooled primeval fireball, or proposing that the universe far beyond the event horizon is steady state but, by a peculiar accident, we live in a kind of expanding bubble, a violent pimple in a much vaster but more placid universe. This idea has the advantage or flaw, depending on your point of view, of being impossible to disprove by any conceivable experiment, and virtually all cosmologists have abandoned the Steady State hypothesis.

If the universe is not in a steady state, then it is changing, and such changing universes are described by evolutionary cosmologies. They begin in one state, and they end in another. What are the possible fates of the universe in evolutionary cosmologies? If the universe continues to expand at its present rate and galaxies continue to disappear over the event horizon, there will eventually be less and less matter in the visible universe. The distances between galaxies will increase, and there will be fewer and fewer of the spiral nebulae for the successors of Slipher, Hubbell and Humason to view. Eventually the distance from our Galaxy to the nearest galaxy will exceed the distance to the event horizon, and astronomers will no longer be able to see even the nearest galaxy except in (very) old books and photographs. Because of the gravity that holds the stars in our Galaxy together, the expanding universe will not dissipate our Galaxy, but even here a strange and desolate fate awaits us. For one thing, the stars are evolving, and in tens or hundreds of billions of years most present stars will have become small and dark dwarf stars. The remainder will have collapsed to neutron stars or black holes. No new matter will be available for a vigorous younger generation of stars. The Sun, the stars, the entire Milky Way Galaxy, will slowly turn off. The lights in the night sky will go out.

But in such a universe there is a further evolution still. We are used to the idea of radioactive elements, certain kinds of atoms that spontaneously decay or fall to pieces. Ordinary uranium is one example. But we are less familiar with the idea that every atom except iron is radioactive, given a long enough period of time. Even the most stable atoms will radioactively decay, emit alpha and other particles, and fall to pieces, leaving only iron, if we wait long enough. How long? The American physicist Freeman Dyson of the Institute for Advanced Study calculates that the half-life of iron is about 10^{500} years, a one followed by five hundred zeros—a number so large that it would take a dedicated numerologist the better part of ten minutes just to write it down. So if we wait just a little longer—10^{600} years would do just fine—not only would the stars have gone out, but all the matter in the universe not in neutron stars or black holes would have decayed into the ultimate nuclear dust. Eventually, galaxies will have vanished altogether. Suns will have blackened, matter disintegrated, and no conceivable possibility will remain for the survival of life or intelligence or civilizations—a cold and dark and desolate death of the universe.

But need the universe expand forever? If I stand on a small asteroid and throw a rock up, it will leave the asteroid, there being on such a worldlet not enough gravity to drag the rock back. If I throw the same rock at the same speed from the surface of the Earth, it will of course turn around and fall down because of the substantial gravity of our planet. But the same sort of physics applies to the universe as a whole. If there is less than a certain amount of matter, each galaxy will feel an insufficient tug from the gravitational attraction of the others to be slowed down appreciably, and the expansion of the universe will continue forever. On the other hand, if there is more than a certain critical mass, the expansion will eventually slow, and we will be saved from the desolation teleology of a universe that expands forever.

What, then, would be the fate of the universe? Why, then an observer would see expansion eventually replaced by contraction, the galaxies slowly and then at an ever-increasing pace approaching one another, a careening, devastating smashing together of galaxies, worlds, life, civilizations and matter until every structure in the universe is utterly destroyed and all the matter in the cosmos converted into energy: instead of a universe ending in cold and tenuous desolation, a universe finishing in a hot and dense fireball. It is very likely that such a fireball would rebound, leading to a new expansion of the universe and, if the laws of nature remain the same, a new incarnation of matter, a new set of condensations of galaxies and stars and planets, a new evolution of life and intelligence. But information from our universe would not trickle into that next one and, from our vantage point, such an oscillating cosmology is as definitive and depressing an end as the expansion that never stops.

The distinction between a Big Bang with expansion forever and an Oscillating Universe clearly turns on the amount of matter there is. If the critical amount of matter is exceeded, we live in an Oscillating Universe. Otherwise we live in one that expands forever. The expansion times—measured in tens of billions of years—are so long that these cosmological issues do not affect any immediate human concerns. But they are of the most profound import for our view of the nature and fate of the universe and—only a little more remotely—of ourselves.

In a remarkable scientific paper published in the December 15, 1974 issue of the *Astrophysical Journal*, a wide range of observational evidence is brought to bear on the question of whether the universe will continue to expand forever (an "open" universe) or whether it will gradually slow down and recontract (a "closed" universe), perhaps as part of an infinite series of oscillations. The work is by J. Richard Gott III and James E. Gunn, then both of the California Institute of Technology, and David N. Schramm and Beatrice M. Tinsely, then of the University of Texas. In one of their arguments they review calculations of the amount of mass in and between galaxies in "nearby" well-observed regions of space and extrapolate to the rest of the universe: they find that there is not enough matter to slow the expansion down.

Ordinary hydrogen has a nucleus comprising a single proton. Heavy hydrogen, called deuterium, has a nucleus comprising one proton and one neutron. An astronomical telescope in Earth orbit called "Copernicus" has measured, for the first time, the amount of deuterium between the stars. Deuterium must have been made in the Big Bang in an amount that depends on the early density of the universe. The early density of the universe is connected with the present density of the universe. The amount of deuterium found by "Copernicus" implies a value to the early density of the universe and suggests that the present density is insufficient to prevent the universe from expanding forever.* And what is said to be the best value of the Hubbell constant—which specifies how much faster more distant galaxies are receding from us than nearby ones—is consistent with this whole story.

Gott and his colleagues stressed that there may be loopholes in their argument, that it may be possible to hide intergalactic matter in ways which we could not then detect. Evidence for such missing mass has now begun to emerge. The High Energy Astronomical Observatories (HEAO) are a set of satellites orbiting the Earth and scanning the universe for particles and radiation that we cannot detect down here, under our thick

* But there is still a debate on how much deuterium can be made in the hot insides of stars and later spewed back into the interstellar gas. If this is substantial, the present deuterium abundance will have less impact on the density of the early universe.

blanket of air. Satellites of this sort have detected intense emission of X-rays from clusters of galaxies, from intergalactic spaces where there was hitherto no hint of any matter. Extremely hot gas between the galaxies would be invisible to other experimental methods and therefore missed in the inventory of cosmic matter made by Gott and his colleagues. What is more, ground-based radio astronomical studies with the Arecibo Observatory in Puerto Rico have shown that the matter in galaxies extends far beyond the optical light from the apparent edges of galaxies. When we look at a photograph of a galaxy, we see an edge or periphery beyond which there is no apparent luminous matter. But the Arecibo radio telescope has found that the matter fades off extremely slowly and that there is substantial dark matter in the peripheries and exteriors of galaxies, which had been missed by previous surveys.

The amount of missing matter required to make the universe ultimately collapse is substantial. It is thirty times the matter in standard inventories such as Gott's. But it may be that the dark gas and dust in the galactic outskirts, and the astonishingly hot gas glowing in X-rays between the galaxies, together constitute just enough matter to close the universe, prevent an expansion forever—but condemn us to an irrevocable end in a cosmic fireball 50 billion or 100 billion years hence. The issue is still teetering. The deuterium evidence points the other way. Our inventories of mass are still far from complete. But as new observational techniques develop we will have the capability of detecting more and more of any missing mass, and so it would seem that the pendulum is swinging toward a closed universe.

It is a good idea not to make up our minds prematurely on this issue. It is probably best not to let our personal preferences influence the decision. Rather, in the long tradition of successful science, we should permit nature to reveal the truth to us. But the pace of discovery is quickening. The nature of the universe emerging from modern experimental cosmology is very different from that of the ancient Greeks who speculated on the universe and the gods. If we have avoided anthropocentrism, if we have truly and dispassionately considered all alternatives, it may be that in the next few decades we will, for the first time, rigorously determine the nature and fate of the universe. And then we shall see if Gott knows.

CHAPTER 25

THE AMNIOTIC UNIVERSE

◆◆◆

It is as natural to man to die as to be born; and to a little infant, perhaps, the one is as painful as the other.

FRANCIS BACON,
Of Death (1612)

The most beautiful thing we can experience is the mysterious. It is the source of all true art and science. He to whom this emotion is a stranger, who can no longer wonder and stand rapt in awe, is as good as dead: his eyes are closed. . . . To know that what is impenetrable to us really exists, manifesting itself as the highest wisdom and the most radiant beauty which our dull facilities can comprehend only in the most primitive forms—this knowledge, this feeling, is at the center of true religiousness. In this sense, and in this sense only, I belong to the ranks of the devoutly religious men.

ALBERT EINSTEIN,
What I Believe (1930)

WILLIAM WOLCOTT died and went to heaven. Or so it seemed. Before being wheeled to the operating table, he had been reminded that the surgical procedure would entail a certain risk. The operation was a success, but just as the anaesthesia was wearing off his heart went into fibrillation and he died. It seemed to him that he had somehow left his body and was able to look down upon it, withered and pathetic, covered only by a sheet, lying on a hard and unforgiving surface. He was only a little sad, regarded his body one last time—from a great height, it seemed—and continued a kind of upward journey. While his surroundings had been suffused by a strange permeating darkness, he realized that things were now getting brighter—looking up, you might say. And then he was being illuminated from a distance, flooded with light. He entered a kind of radiant kingdom and there, just ahead of him, he could make out in silhouette, magnificently lit from behind, a great godlike figure whom he was now effortlessly approaching. Wolcott strained to make out His face . . .

And then awoke. In the hospital operating room where the defibrillation machine had been rushed to him, he had been resuscitated at the last possible moment. Actually, his heart had stopped, and by some definitions of this poorly understood process, he had died. Wolcott was certain that he *had* died, that he had been vouchsafed a glimpse of life after death and a confirmation of Judaeo-Christian theology.

Similar experiences, now widely documented by physicians and others, have occurred all over the world. These perithanatic, or near-death, epiphanies have been experienced not only by people of conventional Western religiosity but also by Hindus and Buddhists and skeptics. It seems plausible that many of our conventional ideas about heaven are derived from such near-death experiences, which must have been related regularly over the millennia. No news could have been more interesting or more hopeful than that of the traveler returned, the report that there is a voyage and a life after death, that there is a God who awaits us, and that upon death we feel grateful and uplifted, awed and overwhelmed.

For all I know, these experiences may be just what they seem and a vindication of the pious faith that has taken such a pummeling from science in the past few centuries. Personally, I would be delighted if there were a life after death—especially if it permitted me to continue to learn about this world and others, if it gave me a chance to discover how history turns out. But I am also a scientist, so I think about what other explanations are possible. How could it be that people of all ages, cultures and eschatological predispositions have the *same sort* of near-death experience?

We know that similar experiences can be induced with fair regularity, cross-culturally, by psychedelic drugs.* Out-of-body experiences are induced by dissociative anaesthetics such as the ketamines (2-[o-chlorophenyl]-2-[methylamino] cyclohexanones.) The illusion of flying is induced by atropine and other belladonna alkaloids, and these molecules, obtained, for example, from mandrake or jimson weed, have been used regularly by European witches and North American *curanderos* ("healers") to experience, in the midst of

---

* It is interesting to wonder why psychedelic molecules exist —especially in great abundance—in a variety of plants. The psychedelics are unlikely to provide any immediate benefit for the plant. The hemp plant probably does not get high from its complement of $^{1}\Delta$ tetrahydrocannabinol. But human beings *cultivate* hemp because the hallucinogenic properties of marijuana are widely prized. There is evidence that in some cultures psychedelic plants are the only domesticated vegetation. It is possible that in such ethnobotany a symbiotic relationship has developed between the plants and the humans. Those plants which by accident provide desired psychedelics are preferentially cultivated. Such artificial selection can exert an extremely powerful influence on subsequent evolution in relatively short time periods—say, tens of thousands of years—as is apparent by comparing many domesticated animals with their wild forebears. Recent work also makes it likely that psychedelic substances work because they are close chemical congeners of natural substances, produced by the brain, which inhibit or enhance neural transmission, and which may have among their physiological functions the induction of endogenous changes in perception or mood.

religious ecstasy, soaring and glorious flight. MDA (2,4-methylenedioxyamphetamine) tends to induce age regression, an accessing of experiences from youth and infancy which we had thought entirely forgotten. DMT (N,N-dimethyltryptamine) induces micropsia and macropsia, the sense of the world shrinking or expanding, respectively—a little like what happens to Alice after she obeys instructions on small containers reading "Eat me" or "Drink me." LSD (lysergic acid diethylamide) induces a sense of union with the universe, as in the identification of Brahman with Atman in Hindu religious belief.

Can it really be that the Hindu mystical experience is pre-wired into us, requiring only 200 micrograms of LSD to be made manifest? If something like ketamine is released in times of mortal danger or near-death, and people returning from such an experience always provide the same account of heaven and God, then must there not be a sense in which Western as well as Eastern religions are hard-wired in the neuronal architecture of our brains?

It is difficult to see why evolution should have selected brains that are predisposed to such experiences, since no one seems to die or fail to reproduce from a want of mystic fervor. Might these drug-inducible experiences as well as the near-death epiphany be due merely to some evolutionarily neutral wiring defect in the brain which, by accident, occasionally brings forth altered perceptions of the world? That possibility, it seems to me, is extremely implausible, and perhaps no more than a desperate rationalist attempt to avoid a serious encounter with the mystical.

The only alternative, so far as I can see, is that every human being, without exception, has already shared an experience like that of those travelers who return from the land of death: the sensation of flight; the emergence from darkness into light; an experience in which, at least sometimes, a heroic figure can be dimly perceived, bathed in radiance and glory. There is only one common experience that matches this description. It is called birth.

HIS NAME IS STANISLAV GROF. In some pronunciations his first and last names rhyme. He is a physician and a psychiatrist who has, for more than twenty years, employed LSD and other psychedelic drugs in psychotherapy. His work long antedates the American drug culture, having begun in Prague, Czechoslovakia, in 1956 and continuing in recent years in the slightly different cultural setting of Baltimore, Maryland. Grof probably has more continuing scientific experience on the effects of psychedelic drugs on patients than anyone else.* He stresses that whereas LSD can be used for recreational and aesthetic purposes, it can have other and more profound effects, one of which is the accurate recollection of perinatal experiences. "Perinatal" is a neologism for "around birth," and is intended to apply not just to the time immediately after birth but to the time before as well. (It is a parallel construction to "perithanatic," near-death.) He reports a large number of patients who, after a suitable number of sessions, actually re-experience rather than merely recollect profound experiences, long gone and considered intractable to our imperfect memories, from perinatal times. This is, in fact, a fairly common LSD experience, by no means limited to Grof's patients.

Grof distinguishes four perinatal stages recovered under psychedelic therapy. Stage 1 is the blissful complacency of the child in the womb, free of all anxiety, the center of a small, dark, warm universe—a cosmos in an amniotic sac. In its intrauterine state the fetus seems to experience something very close to the oceanic ecstasy described by Freud as a fount of the religious

* A fascinating description of Grof's work and the entire range of psychedelics can be found in the forthcoming book *Psychedelic Drugs Reconsidered* by Lester Grinspoon and James Bakalar (New York, Basic Books, 1979). Grof's own description of his findings can be found in *Realms of the Human Unconscious* by S. Grof (New York, E. P. Dutton, 1976) and *The Human Encounter with Death* by S. Grof and J. Halifax (New York, E. P. Dutton, 1977).

sensibility. The fetus is, of course, moving. Just before birth it is probably as alert, perhaps even more alert, than just after birth. It does not seem impossible that we may occasionally and imperfectly remember this Edenic, golden age, when every need—for food, oxygen, warmth and waste disposal—was satisfied before it was sensed, provided automatically by a superbly designed life-support system; and, in dim recollection years later, describe it as "being one with the universe."

In Stage 2, the uterine contractions begin. The walls to which the amniotic sac is anchored, the foundation of the stable intrauterine environment, become traitorous. The fetus is dreadfully compressed. The universe seems to pulsate, a benign world suddenly converted into a cosmic torture chamber. The contractions may last intermittently for hours. As time goes on, they become more intense. No hope of surcease is offered. The fetus has done nothing to deserve such a fate, an innocent whose cosmos has turned upon it, administering seemingly endless agony. The severity of this experience is apparent to anyone who has seen a neonatal cranial distortion that is still evident days after birth. While I can understand a strong motivation to obliterate utterly any trace of this agony, might it not resurface under stress? Might not, Grof asks, the hazy and repressed memory of this experience prompt paranoid fantasies and explain our occasional human predilections for sadism and masochism, for an identification of assailant and victim, for that childlike zest for destruction in a world which, for all we know, may tomorrow become terrifyingly unpredictable and unreliable? Grof finds recollections in the next stage connected with images of tidal waves and earthquakes, the analogues in the physical world of the intrauterine betrayal.

Stage 3 is the end of the birth process, when the child's head has penetrated the cervix and might, even if the eyes are closed, perceive a tunnel illuminated at one end and sense the brilliant radiance of the extrauterine world. The discovery of light for a creature that has lived its entire existence in darkness must be a profound and on some level an unforgettable experience.

And there, dimly made out by the low resolution of the newborn's eyes, is some godlike figure surrounded by a halo of light—the Midwife or the Obstetrician or the Father. At the end of a monstrous travail, the baby flies away from the uterine universe, and rises toward the lights and the gods.

Stage 4 is the time immediately after birth when the perinatal apnea has dissipated, when the child is blanketed or swaddled, hugged and given nourishment. If recollected accurately, the contrast between Stages 1 and 2 and 2 and 4, for an infant utterly without other experience, must be very deep and striking; and the importance of Stage 3 as the passage between agony and at least a tender simulacrum of the cosmic unity of Stage 1 must have a powerful influence on the child's later view of the world.

There is, of course, room for skepticism in Grof's account and in my expansion upon it. There are many questions to be answered. Do children born before labor by Caesarean section never recall the agonizing Stage 2? Under psychedelic therapy, do they report fewer images of catastrophic earthquakes and tidal waves than those born by normal deliveries? Conversely, are children born after the particularly severe uterine contractions induced in "elective labor" by the hormone oxytocin* more likely to acquire the psychological burdens of Stage 2? If the mother is given a strong sedative, will the baby upon maturity recall a very different transition from Stage 1 directly to Stage 4 and never report, in a perithanatic experience, a radiant epiphany? Can neonates resolve an image at the moment of birth or are they merely sensitive to light and

* Astonishingly, oxytocin turns out to be an ergot derivative that is chemically related to psychedelics such as LSD. Since it induces labor, it is at least a plausible hypothesis that some similar natural substance is employed by nature to induce uterine contractions. But this would imply some fundamental connection for the mother—and perhaps for the child—between birth and psychedelic drugs. Perhaps it is therefore not so implausible that, much later in life under the influence of a psychedelic drug, we recall the birth experience—the event during which we first experienced psychedelic drugs.

darkness? Might the description, in the near-death experience, of a fuzzy and glowing god without hard edges be a perfect recollection of an imperfect neonatal image? Are Grof's patients selected from the widest possible range of human beings or are these accounts restricted to an unrepresentative subset of the human community?

It is easy to understand that there might be more personal objections to these ideas, a resistance perhaps similar to the kind of chauvinism that can be detected in justifications of carnivorous eating habits: the lobsters have no central nervous system; they don't mind being dropped alive into boiling water. Well, maybe. But the lobster-eaters have a vested interest in this particular hypothesis on the neurophysiology of pain. In the same way I wonder if most adults do not have a vested interest in believing that infants possess very limited powers of perception and memory, that there is no way the birth experience could have a profound and, in particular, a profoundly negative influence.

If Grof is right about all this, we must ask why such recollections are possible—why, if the perinatal experience has produced enormous unhappiness, evolution has not selected out the negative psychological consequences. There are some things that newborn infants must do. They must be good at sucking; otherwise they will die. They must, by and large, look cute because at least in previous epochs of human history, infants who in some way seemed appealing were better taken care of. But *must* newborn babies see images of their environment? *Must* they remember the horrors of the perinatal experience? In what sense is there survival value in that? The answer might be that the pros outweigh the cons—perhaps the loss of a universe to which we are perfectly adjusted motivates us powerfully to change the world and improve the human circumstance. Perhaps that striving, questing aspect of the human spirit would be absent if it were not for the horrors of birth.

I am fascinated by the point—which I stress in my book *The Dragons of Eden*—that the pain of childbirth is especially marked in human mothers because of the enormous recent growth of the brain in the last few million years. It would seem that our intelligence is the source of our unhappiness in an almost literal way; but it would also imply that our unhappiness is the source of our strength as a species.

These ideas may cast some light on the origin and nature of religion. Most Western religions long for a life after death; Eastern religions for relief from an extended cycle of deaths and rebirths. But both promise a heaven or satori, an idyllic reunion of the individual and the universe, a return to Stage 1. Every birth is a death—the child leaves the amniotic world. But devotees of reincarnation claim that every death is a birth—a proposition that could have been triggered by perithanatic experiences in which the perinatal memory was recognized as a recollection of birth. ("There was a faint rap on the coffin. We opened it, and it turned out that Abdul had not died. He had awakened from a long illness which had cast its spell upon him, and he told a strange story of being born once again.")

Might not the Western fascination with punishment and redemption be a poignant attempt to make sense of perinatal Stage 2? Is it not better to be punished for something—no matter how implausible, such as original sin—than for nothing? And Stage 3 looks very much like a common experience, shared by all human beings, implanted into our earliest memories and occasionally retrieved in such religious epiphanies as the near-death experience. It is tempting to try to understand other puzzling religious motifs in these terms. *In utero* we know virtually nothing. In Stage 2 the fetus gains experience of what might very well in later life be called evil—and then is forced to leave the uterus. This is entrancingly close to eating the fruit of the tree of the knowledge of good and evil and then experiencing the "expulsion" from Eden.* In Michelangelo's famous painting on the ceiling of the Sistine Chapel, is the finger of God an

* A different but not inconsistent hypothesis on the Eden metaphor, in phylogeny rather than ontogeny, is described in *The Dragons of Eden.*

obstetrical finger? Why is baptism, especially total-immersion baptism, widely considered a symbolic rebirth? Is holy water à metaphor for amniotic fluid? Is not the entire concept of baptism and the "born again" experience an explicit acknowledgment of the connection between birth and mystical religiosity?

If we study some of the thousands of religions on the planet Earth, we are impressed by their diversity. At least some of them seem stupefyingly harebrained. In doctrinal details, mutual agreement is rare. But many great and good men and women have stated that behind the apparent divergences is a fundamental and important unity; beneath the doctrinal idiocies is a basic and essential truth. There are two very different approaches to a consideration of tenets of belief. On the one hand, there are the believers, who are often credulous, and who accept a received religion literally, even though it may have internal inconsistencies or be in strong variance with what we know reliably about the external world or ourselves. On the other hand, there are the stern skeptics, who find the whole business a farrago of weak-minded nonsense. Some who consider themselves sober rationalists resist even considering the enormous corpus of recorded religious experience. These mystical insights must mean something. But what? Human beings are, by and large, intelligent and creative, good at figuring things out. If religions are fundamentally silly, why is it that so many people believe in them?

Certainly, bureaucratic religions have throughout human history allied themselves with the secular authorities, and it has frequently been to the benefit of those ruling a nation to inculcate the faith. In India, when the Brahmans wished to keep the "untouchables" in slavery, they proffered divine justification. The same self-serving argument was employed by whites, who actually described themselves as Christians, in the ante-bellum American South to support the enslavement of blacks. The ancient Hebrews cited God's direction and encouragement in the random pillage and murder they sometimes visited on innocent peoples. In medieval times the Church held out the hope of a glorious life after death to those upon whom it urged contentment with their lowly and impoverished station. These examples can be multiplied indefinitely, to include virtually all the world's religions. We can understand why the oligarchy might favor religion when, as is often the case, religion justifies oppression—as Plato, a dedicated advocate of book-burning, did in the *Republic*. But why do the oppressed so eagerly go along with these theocratic doctrines?

The general acceptance of religious ideas, it seems to me, can only be because there is something in them that resonates with our own certain knowledge—something deep and wistful; something every person recognizes as central to our being. And that common thread, I propose, is birth. Religion is fundamentally mystical, the gods inscrutable, the tenets appealing but unsound because, I suggest, blurred perceptions and vague premonitions are the best that the newborn infant can manage. I think that the mystical core of the religious experience is neither literally true nor perniciously wrong-minded. It is rather a courageous if flawed attempt to make contact with the earliest and most profound experience of our lives. Religious doctrine is fundamentally clouded because not a single person has ever at birth had the skills of recollection and retelling necessary to deliver a coherent account of the event. All successful religions seem at their nucleus to make an unstated and perhaps even unconscious resonance with the perinatal experience. Perhaps when secular influences are subtracted, it will emerge that the most successful religions are those which perform this resonance best.

Attempts at rationalistic explanations of religious belief have been resisted vigorously. Voltaire argued that if God did not exist Man would be obliged to invent him, and was reviled for the remark. Freud proposed that a paternalistic God is partly our projection as adults of our perceptions of our fathers when we were infants; he also called his book on religion *The Future of an Illusion*. He was not despised as much as we might imagine for these views, but perhaps only because he had already demonstrated his disreputability by introducing such scandalous notions as infantile sexuality.

Why is the opposition to rational discourse and reasoned argument in religion so strong? In part, I think it is because our common perinatal experiences are real but resist accurate recollection. But another reason, I think, has to do with the fear of death. Human beings and our immediate ancestors and collateral relatives, such as the Neanderthals, are probably the first organisms on this planet to have a clear awareness of the inevitability of our own end. We will die and we fear death. This fear is worldwide and transcultural. It probably has significant survival value. Those who wish to postpone or avoid death can improve the world, reduce its perils, make children who will live after us, and create great works by which they will be remembered. Those who propose rational and skeptical discourse on things religious are perceived as challenging the remaining widely held solution to the human fear of death, the hypothesis that the soul lives on after the body's demise.* Since we feel strongly, most of us, about wishing not to die, we are made uncomfortable by those who suggest that death is the end; that the personality and the soul of each of us will not live on. But the soul hypothesis and the God hypothesis are separable; indeed, there are some human cultures in which the one can be found without the other. In any case, we do not advance the human cause by refusing to consider ideas that make us frightened.

---

* One curious variant is given in Arthur Schnitzler's *Flight Into Darkness:* ". . . at all the moments of death of any nature, one lives over again his past life with a rapidity inconceivable to others. This remembered life must also have a last moment, and this last moment its own last moment, and so on, and hence, dying is itself eternity, and hence, in accordance with the theory of limits, one may approach death but can never reach it." In fact, the sum of an infinite series of this sort is finite, and the argument fails for mathematical as well as other reasons. But it is a useful reminder that we are often willing to accept desperate measures to avoid a serious confrontation with the inevitability of death.

Those who raise questions about the God hypothesis and the soul hypothesis are by no means all atheists. An atheist is someone who is certain that God does not exist, someone who has compelling evidence against the existence of God. I know of no such compelling evidence. Because God can be relegated to remote times and places and to ultimate causes, we would have to know a great deal more about the universe than we do now to be sure that no such God exists. To be certain of the existence of God and to be certain of the nonexistence of God seem to me to be the confident extremes in a subject so riddled with doubt and uncertainty as to inspire very little confidence indeed. A wide range of intermediate positions seems admissible, and considering the enormous emotional energies with which the subject is invested, a questing, courageous and open mind seems to be the essential tool for narrowing the range of our collective ignorance on the subject of the existence of God.

When I give lectures on borderline or pseudo or folk science (along the lines of Chapters 5 through 8 of this book) I am sometimes asked if similar criticism should not be applied to religious doctrine. My answer is, of course, yes. Freedom of religion, one of the rocks upon which the United States was founded, is essential for free inquiry. But it does not carry with it any immunity from criticism or reinterpretation for the religions themselves. The words "question" and "quest" are cognates. Only through inquiry can we discover truth. I do not insist that these connections between religion and perinatal experience are correct or original. Many of them are at least implicit in the ideas of Stanislav Grof and the psychoanalytic school of psychiatry, particularly Otto Rank, Sandor Ferenczi and Sigmund Freud. But they are worth thinking about.

There is, of course, a great deal more to the origin of religion than these simple ideas suggest. I do not propose that theology is physiology entirely. But it would be astonishing, assuming we really can remember our perinatal experiences, if they did not affect in the deepest way our attitudes on birth and death, sex and childhood, on purpose and ethics, on causality and God.

AND COSMOLOGY. Astronomers studying the nature and origin and fate of the universe make elaborate observations, describe the cosmos in differential equations and the tensor calculus, examine the universe from X-rays to radio waves, count the galaxies and determine their motions and distances—and when all is done a choice is to be made between three different views: a Steady State cosmology, blissful and quiet; an Oscillating Universe, in which the universe expands and contracts, painfully and forever; and a Big Bang expanding universe, in which the cosmos is created in a violent event, suffused with radiation ("Let there be light") and then grows and cools, evolves and becomes quiescent, as we saw in the previous chapter. But these three cosmologies resemble with an awkward, almost embarrassing precision the human perinatal experiences of Grof's Stages 1, 2, and 3 plus 4, respectively.

It is easy for modern astronomers to make fun of the cosmologies of other cultures—for example, the Dogon idea that the universe was hatched from a cosmic egg (Chapter 6). But in light of the ideas just presented, I intend to be much more circumspect in my attitudes toward folk cosmologies; their anthropocentrism is just a little bit easier to discern than ours. Might the puzzling Babylonian and Biblical references to waters above and below the firmament, which Thomas Aquinas struggled so painfully to reconcile with Aristotelian physics, be merely an amniotic metaphor? Are we incapable of constructing a cosmology that is not some mathematical encrypting of our own personal origins?

Einstein's equations of general relativity admit a solution in which the universe expands. But Einstein, inexplicably, overlooked such a solution and opted for an absolutely static, nonevolving cosmos. Is it too much to inquire whether this oversight had perinatal rather than mathematical origins? There is a demonstrated reluctance of physicists and astronomers to accept Big Bang cosmologies in which the universe expands forever, although conventional Western theologians are more or less delighted with the prospect. Might this dispute, based almost certainly on psychological predispositions, be understood in Grofian terms?

I do not know how close the analogies are between personal perinatal experiences and particular cosmological models. I suppose it is too much to hope that the originators of the Steady State hypothesis were each born by Caesarean section. But the analogies are very close, and the possible connection between psychiatry and cosmology seems very real. Can it really be that every possible mode of origin and evolution of the universe corresponds to a human perinatal experience? Are we such limited creatures that we are unable to construct a cosmology that differs significantly from one of the perinatal stages?* Is our ability to know the universe hopelessly ensnared and enmired in the experiences of birth and infancy? Are we doomed to recapitulate our origins in a pretense of understanding the universe? Or might the emerging observational evidence gradually force us into an accommodation with and an understanding of that vast and awesome universe in which we float, lost and brave and questing?

It is customary in the world's religions to describe Earth as our mother and the sky as our father. This is true of Uranus and Gaea in Greek mythology, and also among Native Americans, Africans, Polynesians, indeed most of the peoples of the planet Earth. However, the very point of the perinatal experience is that we leave our mothers. We do it first at birth and then again

---

* Kangaroos are born when they are little more than embryos and must then make, entirely unassisted, a heroic journey hand over hand from birth canal to pouch. Many fail this demanding test. Those who succeed find themselves once again in a warm, dark and protective environment, this one equipped with teats. Would the religion of a species of intelligent marsupials invoke a stern and implacable god who severely tests marsupialkind? Would marsupial cosmology deduce a brief interlude of radiation in a premature Big Bang followed by a "Second Dark," and then a much more placid emergence into the universe we know?

when we set out into the world by ourselves. As painful as those leave-takings are, they are essential for the continuance of the human species. Might this fact have some bearing on the almost mystical appeal that space flight has, at least for many of us? Is it not a leaving of Mother Earth, the world of our origins, to seek our fortune among the stars? This is precisely the final visual metaphor of the film *2001: A Space Odyssey*. Konstantin Tsiolkovsky was a Russian schoolmaster, almost entirely self-educated, who, around the turn of the century, formulated many of the theoretical steps that have since been taken in the development of rocket propulsion and space flight. Tsiolkovsky wrote: "The Earth is the cradle of mankind. But one does not live in the cradle forever."

We are set irrevocably, I believe, on a path that will take us to the stars—unless in some monstrous capitulation to stupidity and greed, we destroy ourselves first. And out there in the depths of space, it seems very likely that, sooner or later, we will find other intelligent beings. Some of them will be less advanced than we; some, probably most, will be more. Will all the space-faring beings, I wonder, be creatures whose births are painful? The beings more advanced than we will have capabilities far beyond our understanding. In some very real sense they will appear to us as godlike. There will be a great deal of growing up required of the infant human species. Perhaps our descendants in those remote times will look back on us, on the long and wandering journey the human race will have taken from its dimly remembered origins on the distant planet Earth, and recollect our personal and collective histories, our romance with science and religion, with clarity and understanding and love.

REFERENCES

CHAPTER 3
THAT WORLD WHICH BECKONS LIKE A LIBERATION

FEUER, LEWIS S., *Einstein and the Generations of Science.* New York, Basic Books, 1974.

FRANK, PHILIPP, *Einstein: His Life and Times.* New York, Knopf, 1953.

HOFFMAN, BANESH, *Albert Einstein: Creator and Rebel.* New York, New American Library, 1972.

SCHILPP, PAUL, ed., *Albert Einstein: Philosopher Scientist.* New York, Tudor, 1951.

CHAPTER 5
NIGHT WALKERS AND MYSTERY MONGERS

"Alexander the Oracle-Monger," in *The Works of Lucian of Samosata.* Oxford, Clarendon Press, 1905.

CHRISTOPHER, MILBOURNE, *ESP, Seers and Psychics.* New York, Crowell, 1970.

COHEN, MORRIS, and NAGEL, ERNEST, *An Introduction to Logic and Scientific Method.* New York, Harcourt Brace, 1934.

EVANS, BERGEN, *The Natural History of Nonsense.* New York, Knopf, 1946.

GARDNER, MARTIN, *Fads and Fallacies in the Name of Science.* New York, Dover, 1957.

MACKAY, CHARLES, *Extraordinary Popular Delusions and the Madness of Crowds.* New York, Farrar, Straus & Giroux, Noonday Press, 1970.

CHAPTER 7
VENUS AND DR. VELIKOVSKY

BRANDT, J. C., MARAN, S. P., WILLIAMSON, R., HARRINGTON, R., COCHRAN, C., KENNEDY, M., KENNEDY, W., and CHAMBERLAIN, V., "Possible Rock Art Records of the Crab Nebula Supernova in the Western United States." *Archaeoastronomy in Pre-Columbian America,* A. F. Aveni, ed. Austin, University of Texas Press, 1974.

BRANDT, J. C., MARAN, S. P., and STECHER, T. P., "Astronomers Ask Archaeologists Aid." *Archaeology,* 21: 360 (1971).

BROWN, H., "Rare Gases and the Formation of the Earth's Atmosphere," in Kuiper (1949).

CAMPBELL, J., *The Mythic Image.* Princeton, Princeton University Press, 1974. (Second printing with corrections, 1975.)

CONNES, P., CONNES, J., BENEDICT, W. S., and KAPLAN, L. D., "Traces of HC*l* and HF in the Atmosphere of Venus." *Ap. J.,* 147: 1230 (1967).

COVEY, C., *Anthropological Journal of Canada,* 13: 2–10 (1975).

DE CAMP, L. S., *Lost Continents: The Atlantis Theme.* New York, Ballantine Books, 1975.

DODD, EDWARD, *Polynesian Seafaring.* New York, Dodd, Mead, 1972.

EHRLICH, MAX, *The Big Eye.* New York, Doubleday, 1949.

GALANOPOULOS, ANGELOS G., "Die ägyptischen Plagen und der Auszug Israels aus geologischer Sicht." *Das Altertum,* 10: 131–137 (1964).

GOULD, S. J., "Velikovsky in Collision." *Natural History* (March 1975), 20–26.

KUIPER, G. P., ed., *The Atmospheres of the Earth and Planets,* 1st ed. Chicago, University of Chicago Press, 1949.

LEACH, E. R., "Primitive Time Reckoning," in *The History of Technology,* edited by C. Singer, E. J. Holmyard, and Hall, A. R. London, Oxford University Press, 1954.

LECAR, M., and FRANKLIN, F., "On the Original Distribution of the Asteroids." *Icarus,* 20: 422–436 (1973).

MAROV, M. YA., "Venus: A Perspective at the Beginning of Planetary Exploration." *Icarus,* 16: 415–461 (1972).

MAROV, M. YA., AVDUEVSKY, V., BORODIN, N., EKONOMOV, A., KERZHANOVICH, V., LYSOV, V., MOSHKIN, B., ROZHDESTVENSKY, M., and RYABOV, O., "Preliminary Results on the Venus Atmosphere from the Venera 8 Descent Module." *Icarus*, 20: 407–421 (1973).

MEEUS, J., "Comments on *The Jupiter Effect*." *Icarus*, 26: 257–267 (1975).

NEUGEBAUER, O., "Ancient Mathematics and Astronomy," in *The History of Technology*, edited by C. Singer, E. J. Holmyard, and Hall, A. R. London, Oxford University Press, 1954.

ÖPIK, ERNST J., "Collision Probabilities with the Planets and the Distribution of Interplanetary Matter." *Proceedings of the Royal Irish Academy*, Vol. 54 (1951), 165–199.

OWEN, T. C., and SAGAN, C., "Minor Constituents in Planetary Atmospheres: Ultraviolet Spectroscopy from the Orbiting Astronomical Observatory." *Icarus*, 16: 557–568 (1972).

POLLACK, J. B., "A Nongray CO_2-H_2O Greenhouse Model of Venus." *Icarus*, 10: 314–341 (1969).

POLLACK, J. B., ERICKSON, E., WITTEBORN, F., CHACKERIAN, C., SUMMERS, A., AUGASON, G., and CAROFF, L., "Aircraft Observation of Venus' Near-infrared Reflection Spectrum: Implications for Cloud Composition." *Icarus*, 23: 8–26 (1974).

SAGAN, C., "The Radiation Balance of Venus." California Institute of Technology, Jet Propulsion Laboratory, Technical Report 32–34, 1960.

———, "The Planet Venus." *Science*, 133: 849 (1961).

———, *The Cosmic Connection*. New York, Doubleday, 1973.

———, "Erosion of the Rocks of Venus." *Nature*, 261:31 (1976).

SAGAN, C., and PAGE T., eds., *UFOs: A Scientific Debate*. Ithaca, N. Y., Cornell University Press, 1973; New York, Norton, 1974.

SILL, G., "Sulfuric Acid in the Venus Clouds." Communications Lunar Planet Lab., University of Arizona, 9: 191–198 (1972).

SPITZER, LYMAN, and BAADE, WALTER, "Stellar Populations and Collisions of Galaxies." *Ap. J.*, 113: 413 (1951).

UREY, H. C., "Cometary Collisions and Geological Periods." *Nature*, 242: 32–33 (1973).

———, *The Planets*. New Haven, Yale University Press, 1951.

VELIKOVSKY, I., *Worlds in Collision*. New York, Dell, 1965. (First printing, Doubleday, 1950.)

———, "Venus, a Youthful Planet." *Yale Scientific Magazine,* 41: 8–11 (1967).

VITALIANO, DOROTHY B., *Legends of the Earth: Their Geologic Origins*. Bloomington, Indiana University Press, 1973.

WILDT, R., "Note on the Surface Temperature of Venus." *Ap. J.*, 91: 266 (1940).

———, "On the Chemistry of the Atmosphere of Venus." *Ap. J.*, 96: 312–314 (1942).

YOUNG, A. T., "Are the Clouds of Venus Sulfuric Acid?" *Icarus,* 18: 564–582 (1973).

YOUNG, L. D. G., and YOUNG, A. T., Comments on "The Composition of the Venus Cloud Tops in Light of Recent Spectroscopic Data." *Ap. J.*, 179: L39 (1973).

APPENDICES TO CHAPTER 7

APPENDIX 1

Simple Collision Physics Discussion of the Probability of a Recent Collision with the Earth by a Massive Member of the Solar System

WE HERE CONSIDER the probability that a massive object of the sort considered by Velikovsky to be ejected from Jupiter might impact Earth. Velikovsky proposes that a grazing or near-collision occurred between this comet and the Earth. We will subsume this idea under the designation "collision" below. Consider a spherical object of radius R moving among other objects of similar size. Collision will occur when the centers of the objects are 2R distant. We may then speak of an effective collision cross section of $\sigma = \pi(2R)^2 = 4\pi R^2$; this is the target area which the center of the moving object must strike in order for a collision to occur. Let us assume that only one such object (Velikovsky's comet) is moving and that the others (the planets in the inner solar system) are stationary. This neglect of the motion of the planets of the inner solar system can be shown to introduce errors smaller than a factor of 2. Let the comet be moving at a velocity v and let the space density of potential targets (the planets of the inner solar system) be n. We will use units in which R is in centimeters (cm), σ is in cm^2, v is in cm/sec, and n is in planets per cm^3; n is obviously a very small number.

While comets have a wide range of orbital inclinations to the ecliptic plane, we will be making the most generous assumptions for Velikovsky's hypothesis if we assume the smallest plausible value for this inclination. If there were no restriction on the orbital inclination of the comet, it would have equal likelihood of moving anywhere in a volume centered on the Sun and of radius r = 5 astronomical units (1 a.u. = 1.5×10^{13} cm), the semi-major axis of the orbit of Jupiter. The larger the volume in which the comet can move, the less likely is any collision of it with another object. Because of Jupiter's rapid rotation, any object flung out from its interior will have a tendency to move in the planet's equatorial plane, which is inclined by 1.2° to the plane of the Earth's revolution about the Sun. However, for the comet to reach the inner part of the solar system at all, the ejection event must be sufficiently energetic that virtually any value of its orbital inclination, i, is plausible. A generous lower limit is then i = 1.2°. We therefore consider the comet to move (see diagram) in an orbit contained somewhere in a wedge-shaped volume, centered on the Sun (the comet's orbit must have the Sun at one focus), and of half-angle i. Its volume is then $(4/3)\pi r^3 \sin i = 4 \times 10^{40}$ cm^3, only 2 percent the full volume of a sphere of radius *r*. Since in that volume there are (disregarding the asteroids) three or four planets, the space density of targets relevant for our problem is about 10^{-40} planets/cm^3. A typical relative velocity of a comet or other object moving on an eccentric orbit in the inner solar system might be about 20 km/sec. The radius of the Earth is R = 6.3×10^8 cm, which is almost exactly the radius of the planet Venus as well.

Now let us imagine that the elliptical path of the comet is, in our mind's eye, straightened out, and that it travels for some time T until it impacts a planet. During that time it will have carved out an imaginary tunnel behind it of volume $\sigma v T$ cm^3, and in that volume there must be just one planet.

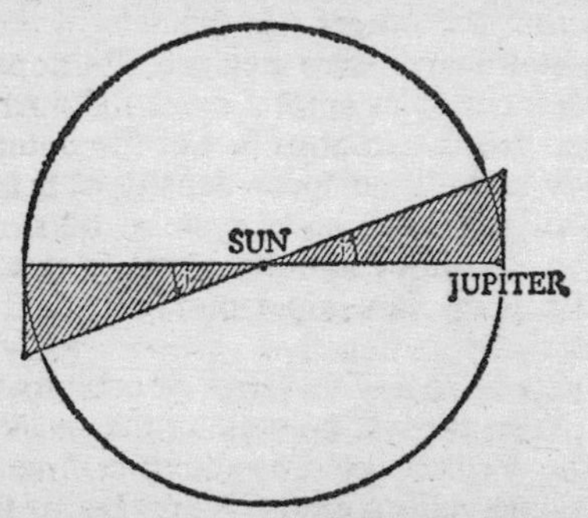

Wedge-shaped volume occupied by Velikovsky's comet.

But $1/n$ is also the volume containing one planet. Therefore, the two quantities are equal and

$$T = (n\sigma v)^{-1};$$

T is called the mean free time.

In reality, of course, the comet will be traveling on an elliptical orbit, and the time for collision will be influenced to some degree by gravitational forces. However, it is easy to show (see, for example, Urey, 1951) that for typical values of v and relatively brief excursions of solar system history such as Velikovsky is considering, the gravitational effects are to increase the effective collision cross section σ by a small quantity, and a rough calculation using the above equation must give approximately the right results.

The objects which have, since the earliest history of the solar system, produced impact craters on the Moon, Earth and the inner planets are ones in highly eccentric orbits: the comets and, especially, the Apollo objects—which are either dead comets or asteroids. Using simple equations for the mean free time, astronomers are able to account to good accuracy for, say, the number of craters on the Moon, Mercury or Mars produced since the formation of these objects: they are the results of the occasional collision of an Apollo object or, more rarely, a comet with the lunar or planetary surface. Likewise, the equation predicts correctly the age of the most recent impact craters on Earth such as Meteor Crater, Arizona. These quantitative agreements between observations and simple collision physics provide some substantial assurance that the same considerations properly apply to the present problem.

We are now able to make some calculations with regard to Velikovsky's fundamental hypothesis. At the present time there are no Apollo objects with diameters larger than a few tens of kilometers. The sizes of objects in the asteroid belt, and indeed anywhere else where collisions determine sizes, are understood by comminution physics. The number of objects in a given size range is proportional to the radius of the object to some negative power, usually in the range of 2 to 4. If, therefore, Velikovsky's proto-Venus comet were a member of some family of objects like the Apollo objects or the comets, the chance of finding one Velikovskian comet 6,000 km in radius would be far less than one-millionth of the chance of finding one some 10 km in radius.

A more probable number is a billion times less likely, but let us give the benefit of the doubt to Velikovsky.

Since there are about ten Apollo objects larger than about 10 km in radius, the chance of there being one Velikovskian comet is then much less than 100,000-to-1 odds against the proposition. The steady-state abundance of such an object would then be (for $r = 4$ a.u., and $i = 1.2°$) $n = (10 \times 10^{-5})/4 \times 10^{40} = 2.5 \times 10^{-45}$ Velikovskian comets/cm^3. The mean free time for collision with Earth would then be $T = 1/(n\sigma v) = 1/[(2.5 \times 10^{-45}\ cm^{-3}) \times (5 \times 10^{18}\ cm^2) \times (2 \times 10^6\ cm\ sec^{-1})] = 4 \times 10^{21}$ secs $\simeq 10^{14}$ years which is much greater than the age of the solar system (5×10^9 years). That is, if the Velikovskian comet were part of the population of other colliding debris in the inner solar system, it would be such a rare object that it would essentially never collide with Earth.

But instead, let us grant Velikovsky's hypothesis for the sake of argument and ask how long his comet would require, after ejection from Jupiter, to collide with a planet in the inner solar system. Then, *n* applies to the abundance of planetary targets rather than Velikovskian comets, and $T = 1/[(10^{-40}\ cm^{-3}) \times (5 \times 10^{18}\ cm^2) \times (2 \times 10^6\ cm\ sec^{-1})] = 10^{15}$ secs $\simeq 3 \times 10^7$ years. Thus, the chance of Velikovsky's "comet" making a single full or grazing collision with Earth within the last few thousand years is $(3 \times 10^4)/(3 \times 10^7) = 10^{-3}$, or one chance in 1,000—if it is independent of the other debris populations. If it is part of such populations, the odds rise to $(3 \times 10^4)/10^{14} = 3 \times 10^{-10}$, or one chance in 3 billion.

A more exact formulation of orbital-collision theory can be found in the classic paper by Ernst Öpik (1951). He considers a target body of mass m_o with orbital elements a_o, $e_o = i_o = 0$ in orbit about a central body of mass M. Then, a test body of mass m with orbital elements a, e, i and period P has a characteristic time T before approaching within distance R of the target body, where

$$\frac{T}{P} \simeq \frac{\pi \sin i |U_x/U|}{Q^2 [1 + 2(m_o + m)/MQU]}$$

$$A = a/a_o,\ Q = R/a_o,$$

$$|U_x| = [2 - A^{-1} - A(1 - e^2)]^{1/2}$$

$$U = \{3 - A^{-1} - 2[A(1 - e^2)]^{1/2} \cos i\}^{1/2};$$

here; U is the relative velocity "at infinity" and U_x is its component along the line of nodes.

If R is taken as the physical radius of the planet, then

	Venus	*Earth*	*Mars*	*Jupiter*
$Q \times 10^5$	5.6	4.3	1.5	8.8
$2m_o/MQ$	0.088	0.14	0.043	21.6

For application of Öpik's results to the present problem, the equations reduce to the following approximation:

$$\frac{T}{P} \simeq \frac{\pi \sin i}{Q^2}$$

Using $P \simeq 5$ years ($a \simeq 3$ a.u.), we have

$$T \simeq 9 \times 10^9 \sin i \text{ years,}$$

or about ⅓ the mean free path lifetime from the simpler argument above.

Note that in both calculations, an approach to within N Earth radii has N^2 times the probability of a physical collision. Thus, for $N = 10$, a miss of 63,000 km, the above values of T must be reduced by two orders of magnitude. This is about 1/6 the distance between the Earth and the Moon.

For the Velikovskian scenario to apply, a closer approach is necessary: the book, after all, is called *Worlds in Collision*. Also, it is claimed (page 72) that, as a result of the passage of Venus by the Earth, the oceans were piled to a height of 1,600 miles. From this it is easy to calculate backwards from simple tidal theory (the tide height is proportional to M/r^3, where M is the mass of Venus and r the distance between the planets during the encounter) that Velikovsky is talking about a grazing collision: the surfaces of Earth and Venus scrape! But note that even a 63,000-km miss does not extricate the hypothesis from the collision physics problems as outlined in this appendix.

Finally, we observe that an orbit which intersects those of Jupiter and Earth implies a high probability of a close reapproach to Jupiter which would eject the object from the solar system before a near-encounter with Earth—a natural example of the trajectory of the Pioneer 10 spacecraft. Therefore, the present existence of the planet Venus must imply that the Velikovskian comet made few subsequent passages to Jupiter, and therefore that its orbit was circularized rapidly. (That there seems to be no way to accomplish such rapid circularization is discussed in the text.) Accordingly, Velikovsky must suppose that the comet's close encounter with Earth occurred soon after its ejection from Jupiter—consistent with the above calculations.

The probability, then, that the comet would have impacted the Earth only some tens of years after its ejection from Jupiter is between one chance in 1 million and one chance in 3 trillion, on the two assumptions on membership in existing debris populations. Even if we were to suppose that the comet was ejected from Jupiter as Velikovsky says, and make the unlikely assumption that it has no relation to any other objects which we see in the solar system today—that is, that smaller objects are never ejected from Jupiter—the mean time for it to have impacted Earth would be about 30 million years, inconsistent with his hypothesis by a factor of about 1 million. Even if we let his comet wander about the inner solar system for centuries before approaching the Earth, the statistics are still powerfully against Velikovsky's hypothesis. When we include the fact that Velikovsky believes in several statistically independent collisions in a few hundred years (see text), the net likelihood that his hypothesis is true becomes vanishing small. His repeated planetary encounters would require what might be called *Worlds in Collusion*.

APPENDIX 2

Consequences of a Sudden Deceleration of Earth's Rotation

> Q. Now, Mr. Bryan, have you ever pondered what would have happened to the Earth if it had stood still?
>
> A. No. The God I believe in could have taken care of that, Mr. Darrow.
>
> Q. Don't you know that it would have been converted into a molten mass of matter?
>
> A. You testify to that when you get on the stand. I will give you a chance.
>
> The Scopes Trial, 1925

THE GRAVITATIONAL acceleration which holds us to the Earth's surface has a value of 10^3 cm sec^{-2} = 1 g. A deceleration of $a = 10^{-2}$ g $= 10$ cm sec^{-2} is almost unnoticeable. How much time, τ, would Earth take to stop its rotation if the resulting deceleration were unnoticeable? Earth's equatorial angular velocity is $\Omega = 2\pi/P = 7.3 \times 10^{-5}$ radians/sec; the equatorial linear velocity is $R\Omega = 0.46$ km/sec. Thus, $\tau = R\Omega/a = 4600$ secs, or a little over an hour.

The specific energy of the Earth's rotation is

$$E = \frac{1}{2} I\Omega^2/M \simeq \frac{1}{5}(R\Omega)^2 \simeq 4 \times 10^8 \text{ erg gm}^{-1},$$

where I is the Earth's principal moment of inertia. This is less than the latent heat of fusion for silicates, $L \simeq 4 \times 10^9$ erg gm^{-1}. Thus, Clarence Darrow was wrong about the Earth melting. Nevertheless, he was on the right track: thermal considerations are in fact fatal to the Joshua story. With a typical specific heat capacity of $c_p \simeq 8 \times 10^6$ erg gm^{-1} deg^{-1}, the stopping and restarting of Earth in one day would have imparted an *average* temperature increment of $\Delta T \simeq 2E/c_p \simeq 100°K$, enough to raise the temperature above the normal boiling point of water. It would have been even worse near the surface and at low latitudes; with $v \simeq R\Omega$, $\Delta T \simeq v^2/c_p \simeq 240°K$. It is doubtful that the inhabitants would have failed to notice so dramatic a climatic change. The deceleration might be tolerable if gradual enough, but not the heat.

APPENDIX 3

Present Temperature of Venus If Heated by a Close Passage to the Sun

THE HEATING of Venus by a presumed close passage by the Sun, and the planet's subsequent cooling by radiation to space are central to the Velikovskian thesis. But nowhere does he calculate either the amount of heating or the rate of cooling. However, at least a crude calculation can readily be performed. An object which grazes the solar photosphere must travel at very high velocities if it originates in the outer solar system: 500 km/sec is a typical value at perihelion passage. But the radius of the Sun is 7×10^{10} cm. Therefore a typical time scale for the heating of Velikovsky's comet is $(1.4 \times 10^{11}\text{cm}) / (5 \times 10^7 \text{ cm/sec}) \simeq 3000$ secs, which is less than an hour. The highest temperature the comet could possibly reach because of its close approach to the Sun is 6,000° K, the temperature of the solar photosphere. Velikovsky does not discuss any further sun-grazing events by his comet; subsequently it becomes the planet Venus, and cools to space—events which occupy, say, 3,500 years up to the present. But both heating and cooling occur radiatively, and the physics of both events is controlled in the same way by the Stefan-Boltzmann law of thermodynamics, according to which the amount of heating and the rate of cooling both are proportional to the temperature to the fourth power. Therefore the ratio of the temperature increment experienced by the comet in 3,000 secs of solar heating to its temperature decrement in 3,500 yrs of radiative cooling is $(3 \times 10^3 \text{ secs}/10^{11} \text{ secs})^{1/4} = 0.013$. The present temperature of Venus from this source would then be at most only $6000 \times 0.013 = 79°$ K, or about the temperature at which air freezes. Velikovsky's mechanism cannot keep Venus hot, even with very generous definitions of the word "hot."

The conclusion would not be altered materially were there to have been several close passes, rather than just one, through the solar photosphere. The source of the high temperature of Venus cannot be one or a few heating events, no matter how dramatic. The hot surface requires a continuous source of heat—which could be either endogenous (radioactive heating from the planetary interior) or exogenous (sunlight). It is now evident, as suggested many years ago (see Wildt, 1940; Sagan, 1960), that the latter is the case: it is the present radiation of the Sun, continuously falling on Venus, which is responsible for its high surface temperature.

APPENDIX 4

Magnetic Field Strengths Necessary to Circularize an Eccentric Cometary Orbit

ALTHOUGH VELIKOVSKY has not, we can calculate approximately the order of magnitude of the magnetic field strength necessary to make a significant perturbation on the motion of a comet. The perturbing field might be from a planet, such as Earth or Mars, to which the comet is about to make a close approach, or from the interplanetary magnetic field. For this field to play an important role, its energy density must be comparable to the kinetic energy density of the comet. (We do not even worry about whether the comet has a distribution of charges and fields which will permit it to respond to the imposed field.) Thus, the condition is

$$\frac{B^2}{8\pi} = \frac{\frac{1}{2}\, mv^2}{(4/3)\,\pi R^3} = (\tfrac{1}{2})\, \rho\, v^2$$

where B is the magnetic field strength in gauss, R is the radius of the comet, m its mass, v its velocity and ρ its density. We note that the condition is independent of the mass of the comet. Taking a typical cometary velocity in the inner solar system of about 25 km/sec, and ρ as the density of Venus, about 5 gm/cm^3, we find that a magnetic field strength of over 10 million gauss is required. (A similar value in electrostatic units would apply if the circularization is electrical rather than magnetic.) Earth's equatorial surface field is about 0.5 gauss. The fields of Mars and Venus are less than 0.01 gauss. The Sun's field is several gauss, ranging up to several hundred gauss in sunspots. Jupiter's field as measured by Pioneer 10 is less than 10 gauss. Typical interplanetary fields are 10^{-5} gauss. There is no way to generate anything approaching a 10 megagauss field on a large scale in the solar system. And there is no sign that such a field was ever experienced in the vicinity of Earth. We recall that the magnetic domains of molten rock in the course of refreezing are oriented by the prevailing field. Had Earth experienced, even fairly briefly, a 10 Mg field 3,500 years ago, rock magnetization evidence would show it clearly. It does not.

INDEX

About the Author

CARL SAGAN is as well known for his literary as for his scientific accomplishments. Winner of the Pulitzer Prize (for *The Dragons of Eden*), he has also received the NASA Medals for Exceptional Scientific Achievement and for Distinguished Public Service, and the Joseph Priestley Award "for distinguished contributions to the welfare of mankind." He is David Duncan Professor of Astronomy and Space Sciences and Director of the Laboratory for Planetary Studies at Cornell University, and has played a leading role in the Mariner, Viking and Voyager expeditions to the planets. He has served as Chairman of the Division for Planetary Sciences of the American Astronomical Society, as President of the Planetology Section of the American Geophysical Union and editor in chief of *Icarus*, the principal professional journal of solar system studies. A leader in establishing the high surface temperatures of Venus and in understanding the seasonal changes on Mars, Dr. Sagan was responsible for the Voyager interstellar record, a message about ourselves sent to other civilizations in space (and described in the book *Murmurs of Earth*).